An Introduction to the Locally-Corrected Nyström Method

Synthesis Lectures on Computational Electromagnetics

Editor
Constantine A. Balanis, *Arizona State University*

An Introduction to the Locally-Corrected Nyström Method

Andrew F. Peterson and Malcolm M. Bibby

ISBN: 978-3-031-00582-4   paperback
ISBN: 978-3-031-01710-0   ebook

DOI 10.1007/978-3-031-01710-0

A Publication in the Springer series
SYNTHESIS LECTURES ON COMPUTATIONAL ELECTROMAGNETICS

Lecture #25
Series ISSN
Synthesis Lectures on Computational Electromagnetics
Print 1932-1252   Electronic 1932-1716

An Introduction to the Locally-Corrected Nyström Method

Andrew F. Peterson and Malcolm M. Bibby
Georgia Institute of Technology

SYNTHESIS LECTURES ON COMPUTATIONAL ELECTROMAGNETICS #25

ABSTRACT

This lecture provides a tutorial introduction to the Nyström and locally-corrected Nyström methods when used for the numerical solutions of the common integral equations of two-dimensional electromagnetic fields. These equations exhibit kernel singularities that complicate their numerical solution. Classical and generalized Gaussian quadrature rules are reviewed. The traditional Nyström method is summarized, and applied to the magnetic field equation for illustration. To obtain high order accuracy in the numerical results, the locally-corrected Nyström method is developed and applied to both the electric field and magnetic field equations. In the presence of target edges, where current or charge density singularities occur, the method must be extended through the use of appropriate singular basis functions and special quadrature rules. This extension is also described.

KEYWORDS

integral equations, numerical methods, quadrature, boundary element method, singular basis functions, method of moments, Green's functions

Contents

CHAPTER 1

Introduction

The Nyström and Locally-corrected Nyström (LCN) methods are procedures for obtaining numerical solutions of integral equations. Integral equations find application in many disciplines; our focus is limited to two-dimensional, electromagnetic scattering problems that have a solution expressed as a scalar function of a single variable. The objective of this lecture is to introduce the reader to the Nyström and LCN methods and provide enough detail so that the reader can implement the techniques in software. The targets considered are limited to perfectly conducting cylinders with simple shapes.

1.1   THE NYSTRÖM METHOD

The Nyström method is a relatively old procedure, originating in the year 1930 [1,2]. The essence of the approach is that the problem domain (the surface of the target in an electromagnetic scattering problem) is divided into N pieces or cells, and the integral operator is replaced with a suitable quadrature rule over each cell. The unknown quantity, which is usually the surface current density, is represented by its samples at the quadrature nodes. Thus, if a q-point quadrature rule is used in every cell, there are Nq unknowns to be determined. The integral equation is enforced at nodes (sample points) of the rule, to obtain a system of equations for the samples of the unknown function at the node points. The solution of this system yields an approximation of the current density, from which other quantities of interest (such as the far-zone scattered field) may be readily determined.

Since readers are likely to be familiar with the *method of moments* (MoM) or *boundary element method* (BEM) for solving integral equations, we offer the following observations on some differences between these techniques and the Nyström method. In the MoM/BEM approach, the surface current is represented by an expansion in basis functions, whose unknown coefficients are determined by the process. In the Nyström method, the primary unknowns are samples of the current density. This is actually a minor difference that can be thought of as a change of basis, as explained in Section 4.3. The accuracy of either approach can be improved by using smaller cells to represent the surface (h-refinement), resulting in more unknowns to be determined. Another way of improving the accuracy is to use a better representation of the current density (p-refinement). In the MoM/BEM, p-refinement is accomplished by using higher degree polynomials as basis functions, which usually increases the amount of computation required per entry of the system matrix. In the Nyström approach, p-refinement is accomplished by using a quadrature rule with more points — a change that does not add to the required computation per entry.

The different computational cost is a consequence of the difference between the system of equations associated with the two techniques. In the Nyström method, the system of equations represents the global quadrature rule implementing the integral operator, and the entries of that system are samples of the kernel (with appropriate quadrature weights). In the MoM/BEM approach, entries of the system of equations are the weighted residuals associated with the integral operator, which are integrals of the operator when applied to the basis functions (and perhaps weighted with testing functions). It is far more expensive to compute the weighted residuals than to compute samples of the kernel, and consequently, it is much cheaper to construct the system of equations for the Nyström method than it is to construct the system of equations for the MoM/BEM. The difference in computational cost is accentuated as the polynomial degree of the representation in either approach increases.

The drawback of the classical Nyström method is that it cannot be used directly for integral equations with singular kernels, including the electric field integral equation (EFIE) or the vector integral equations of 3D electromagnetics, because one of the kernel samples in each equation is infinite in that case. Despite this difficulty, the original method was occasionally adapted to electromagnetics problems [3–9]. The adaptation usually involved the construction of a special quadrature rule capable of integrating the singular kernel [3, 4] or the use of a singularity extraction procedure [5–7] to avoid the infinite sample.

1.2   THE LOCALLY-CORRECTED NYSTRÖM METHOD

The popularity of Nyström approaches in computational electromagnetics was enhanced in the late 1990's with the introduction of the *locally-corrected* Nyström (LCN) technique described by Gedney et al. [10], Canino et al. [11], and Kapur and Long [12]. The LCN method alleviates the difficulty with singular kernels by replacing samples of the singular kernel with "corrected" samples that are finite and have been adjusted to produce the correct near fields for some set of source functions, in accordance with an idea proposed by Strain [13]. A number of different research groups [14–28] have studied LCN discretizations and concluded that the LCN is a reasonable alternative to the MoM and BEM for the numerical treatment of electromagnetics equations. Despite the additional computations needed to "locally-correct" the kernel, the LCN method still offers substantially smaller matrix fill costs compared to the MoM, especially for high order representations where the MoM costs can be substantial.

1.3   A LOOK AHEAD

In the following chapters, the Nyström and LCN methods will be described in a tutorial manner and used in the numerical discretization of common integral equations of electromagnetic fields. Since the concept is highly dependent on numerical quadrature, Chapter 2 reviews the development and performance of some classical quadrature rules. Chapter 3 describes the Nyström method applied to smooth structures modeled by the 2D magnetic field integral equation (MFIE), whose bounded kernel makes the equation amenable to classical Nyström discretizations. One potential advantage of a Nyström discretization is that the solution error should dramatically improve as the order of the quadrature rule is increased. Unfortunately, since the MFIE kernel is not analytic, the improved accuracy is not realized. To achieve improved accuracy and what is commonly referred to as "high-order" behavior, the LCN procedure is introduced in Chapter 4 and utilized for both the MFIE and the electric field integral equation (EFIE). For smooth targets, and typical Gaussian quadrature rules, the LCN does yield high-order accuracy in its numerical solutions. However, for targets with corners or edges, additional refinements are necessary. Chapter 5 extends the discussion of quadrature to the development of generalized Gaussian quadrature rules capable of integrating singular functions, such as the current density near edges. These quadrature rules are implemented within LCN approaches in Chapter 6. When used to analyze structures with edges, the enhanced LCN approach does produce high order accuracy, with error levels that decrease at the same rates as the error for smooth structures.

REFERENCES

[1] E. J. Nyström, "Über die praktische Auflüsung von Integral-gleichungen mit Anwendungen auf Randwertaufgaben," *Acta Math.*, vol. 54, pp. 185–204, 1930. DOI: 10.1007/BF02547521

[2] K. E. Atkinson, *A Survey of Numerical Methods for the Solution of Fredholm Integral Equations of the Second Kind*. Philadelphia: SIAM, 1976.

[3] R. Kress, "Numerical solution of boundary integral equations in time-harmonic electromagnetic scattering," *Electromagnetics*, vol. 10, pp. 1–20, 1990. DOI: 10.1080/02726349008908226

[4] R. Kress, "A Nyström method for boundary integral equations in domains with corners," *Numerische Mathematik*, vol. 58, pp. 145–161, 1990. DOI: 10.1007/BF01385616

[5] J. S. Kot, "Solution of thin-wire integral equations by Nyström methods," *Microwave and Optical Technology Letters*, vol. 3, no. 11, pp. 393–396, November 1990. DOI: 10.1002/mop.4650031109

[6] J. S. Kot, "Computer modeling of MM-wave integrated circuit antennas using the Nyström method," *International Conference on Computation in Electromagnetics*, London, UK, pp. 288–291, November 1991.

[7] J. S. Kot, "Application of Nyström methods to electric- and magnetic-field integral equations," *Proceedings of the 1992 URSI International Symposium on Electromagnetic Theory*, Sydney, NSW, pp. 125–127, August 1992.

[8] J. L. Tsalamengas, "Exponentially converging Nyström's methods for systems of singular integral equations with applications to open/closed strip- or slot-loaded 2-D structures," *IEEE Trans. Antennas Propagat.*, vol. 54, pp. 1549–1558, May 2006. DOI: 10.1109/TAP.2006.874348

[9] J. L. Tsalamengas, "Exponentially converging Nyström methods applied to the integral-integrodifferential equations of oblique scattering/hybrid wave propagation in presence of composite dielectric cylinders of arbitrary cross section," *IEEE Trans. Antennas Propagat.*, vol. 55, pp. 3239–3250, November 2007. DOI: 10.1109/TAP.2007.908833

[10] S. D. Gedney, J. Ottusch, P. Petre, J. Visher, and S. Wandzura, "Efficient high-order discretization schemes for integral equation methods," *Digest of the 1997 IEEE Antennas and Propagation International Symposium*, Montreal, CA, pp. 1814–1817, July 1997.

[11] L. F. Canino, J. J. Ottusch, M. A. Stalzer, J. L. Visher, and S. M. Wandzura, "Numerical Solution of the Helmholtz Equation in 2D and 3D Using a High-Order Nyström Discretization," *J. Comp. Physics*, vol. 146, pp. 627–663, 1998. DOI: 10.1006/jcph.1998.6077

[12] S. Kapur and D. E. Long, "High order Nyström schemes for efficient 3-D capacitance extraction," *Digest of Technical Papers of the IEEE/ACM International Conference on Computer-Aided Design (ICCAD 98)*, pp. 178–185, November 1998. DOI: 10.1145/288548.288604

[13] J. Strain, "Locally corrected multidimensional quadrature rules for singular functions," *SIAM J. Scientific Computing*, vol. 16, pp. 992–1017, 1995. DOI: 10.1137/0916058

[14] S. Gedney, "Application of the high-order Nyström scheme to the integral equation solution of electromagnetic interaction problems, *Proceedings of the IEEE International Symposium on Electromagnetic Compatibility*, Washington, DC, pp. 289–294, August 2000.

[15] G. Liu and S. Gedney, "High-order Nyström solution of the volume EFIE for TM-Wave scattering," *Microwave and Optical Technology Letters*, vol. 25, No. 1, pp. 8–11, 2000. DOI: 10.1002/(SICI)1098-2760(20000405)25:1<8::AID-MOP3>3.0.CO;2-U

[16] G. Liu and S. Gedney, "High-order Nyström solution of the volume EFIE for TE-Wave scattering," *Electromagnetics*, vol. 21, No. 1, pp. 1–14, 2001. DOI: 10.1080/02726340117790

[17] A. F. Peterson, "Accuracy of currents produced by the locally-corrected Nyström method and the method of moments when used with higher-order representations," *Applied Computational Electromagnetics Society (ACES) Journal*, vol. 17, pp. 74–83, March 2002.

[18] S. D. Gedney, "On deriving a locally-corrected Nyström scheme from a quadrature sampled moment method," *IEEE Trans. Antennas Propagat.*, vol. 51, pp. 2402–2412, September 2003. DOI: 10.1109/TAP.2003.816305

[19] S. D. Gedney and C.-C. Lu, "High order solution for the electromagnetic scattering by inhomogeneous dielectric bodies," *Radio Science*, vol. 38, 1015, 8 pages, 2003. DOI: 10.1029/2002RS002700

[20] J. L. Fleming, A. W. Wood, and W. D. Wood, "Locally corrected Nyström method for EM scattering by bodies of revolution," *J. Computational Physics*, vol. 196, pp. 41–52, 2004. DOI: 10.1016/j.jcp.2003.10.029

[21] F. Caliskan and A. F. Peterson, "The need for mixed-order representations with the locally-corrected Nyström method," *IEEE Antennas and Wireless Propagation Letters*, vol. 2, pp. 72–73, April 2003. DOI: 10.1109/LAWP.2003.813383

[22] S. D. Gedney, A. Zhu, and C.-C. Lu, "Study of mixed-order basis functions for the locally corrected Nyström method," *IEEE Trans. Antennas Propagat.*, vol. 52, pp. 2996–3004, November 2004. DOI: 10.1109/TAP.2004.835122

[23] A. F. Peterson, "Application of the locally-corrected Nyström method to the EFIE for the linear dipole," *IEEE Trans. Antennas Propagat.*, vol. 52, pp. 603–605, February 2004. DOI: 10.1109/TAP.2004.823955

[24] A. F. Peterson and M. M. Bibby, "High-order numerical solutions of the MFIE for the linear dipole," *IEEE Trans. Antennas Propagat.*, vol. 52, pp. 2684–2691, October 2004. DOI: 10.1109/TAP.2004.834407

[25] M. S. Tong and W. C. Chew, "A higher-order Nyström scheme for electromagnetic scattering by arbitrarily shaped surfaces," *IEEE Antennas and Wireless Propagation Letters*, vol. 4, pp. 277–280, 2005. DOI: 10.1109/LAWP.2005.853000

[26] M. S. Tong and W. C. Chew, "Nyström method with edge condition for electromagnetic scattering by 2D open structures," *Progress in Electromagnetic Research*, vol. 62, pp. 49–68, 2006. DOI: 10.2528/PIER06021901

[27] R. A. Wildman and D. S. Weile, "Mixed-order testing functions on triangular patches for the locally corrected Nyström method," *IEEE Antennas and Wireless Propagation Letters*, vol. 5, pp. 370–372, 2006. DOI: 10.1109/LAWP.2006.881925

[28] V. Rawat and J. P. Webb, "Scattering from dielectric and metallic bodies using a high-order, Nyström, multilevel fast multipole algorithm," *IEEE Trans. Magnetics*, vol. 42, pp. 521–526, April 2006. DOI: 10.1109/TMAG.2006.871389

CHAPTER 2

Classical Quadrature Rules

The numerical evaluation of integrals is historically known by the name *quadrature*, from the Latin *quadratura*, which denotes the division of a region into squares to estimate its area [1,2]. Another term, *cubature*, denotes the similar process of dividing a region into cubes to estimate the volume. A quadrature *rule* is a specific set of weights $\{w_i\}$ and nodes $\{u_i\}$ associated with a summation of the form

$$\int_0^1 f(u)du \cong \sum_{i=1}^N w_i f(u_i) \tag{2.1}$$

that provides an estimate of the integral. The quadrature rule approximates the integral by the sum of N terms. As N is increased, the approximation usually improves in accuracy. *Closed* rules involve nodes that include the end points ($u_1 = 0$ and/or $u_N = 1$ in the above equation), while *open* rules are those for which all node points lie completely inside the interval ($0 < u_i < 1$). While it is possible to generate rules with one or more nodes located outside the original domain of integration, they are undesirable since the integrand may not be defined there. Furthermore, to guard against unfavorable rounding errors associated with subtraction, it is desired that all the weights be positive.

In the following, we will work with rules defined over the *standard unit cell* ($0 \leq u \leq 1$). Integrals defined over a different interval (a, b) can be converted to a unit domain by a change of variable

$$x = a + (b - a)u \tag{2.2}$$

resulting in the equivalence $g(x) = g\{a + (b - a)u\} = f(u)$ and

$$\int_a^b g(x)dx = \int_0^1 g\{a + (b - a)u\}(b - a)du$$
$$= (b - a)\int_0^1 f(u)du \tag{2.3}$$

where we have employed the Jacobian of the transformation

$$\frac{dx}{du} = b - a . \tag{2.4}$$

The earliest quadrature rules, which are known today as Newton-Cotes rules, used equally-spaced nodes. One of the simplest rules of this type is known as trapezoid rule and is based on the approximation of the integrand as a linear polynomial. Higher-order Newton-Cotes rules employ quadratic, cubic, etc. approximations.

2.1   TRAPEZOID RULE [2, 3]

If the domain of integration is divided into equal-sized intervals, the integral

$$I = \int_0^1 f(u)du \tag{2.5}$$

can be estimated under the assumption that the function is treated as a trapezoid over each interval. For a single interval, the approximation is just the area of a trapezoid of unit height, with one base of length $f(0)$ and the other of length $f(1)$:

$$I \cong \frac{1}{2}\{f(0) + f(1)\} . \tag{2.6}$$

For two intervals, we divide the domain into two trapezoids to obtain

$$I \cong \frac{1}{4}\left\{f(0) + 2f\left(\frac{1}{2}\right) + f(1)\right\} . \tag{2.7}$$

It is convenient to consider an even number ($N = 2^{n-1}$) of intervals, for which a general expression can be written as

$$I_{n,1} = \frac{1}{2^n}\{f(0) + f(1)\} + \frac{1}{2^{n-1}}\sum_{i=1}^{N-1} f\left(\frac{i}{2^{n-1}}\right) . \tag{2.8}$$

In the event that the approximation in (2.8) is not accurate enough, the intervals can each be halved, resulting in a formula involving 2^n intervals that can be computed from

$$I_{n+1,1} = \frac{1}{2}I_{n,1} + \frac{1}{2^n}\sum_{i=1}^{N} f\left(\frac{2i-1}{2^n}\right) . \tag{2.9}$$

The expression in (2.9) uses $I_{n,1}$, directly, to avoid re-sampling the integrand at the nodes already employed in (2.8) and thereby saves half the function evaluations that would otherwise be required to produce the estimate for 2^n intervals. (Usually, the function evaluations are the most expensive part of the computation.) Equivalently, for the same number of function evaluations necessary to produce $I_{n+1,1}$, we obtain two different approximations to I. As long as $I_{n+1,1}$ is nonzero, the two approximations can be used to estimate the error in I as

$$E \cong \frac{|I_{n+1,1} - I_{n,1}|}{|I_{n+1,1}|} . \tag{2.10}$$

To evaluate a given integral to a specified accuracy, $I_{0,1}$ and $I_{1,1}$ are computed, and the procedure is continued in an iterative manner by evaluating (2.9) as n is incremented until E is below a desired accuracy.

The weights and nodes of the trapezoid rule are summarized in Table 2.1 for 2–5 nodes.

nodes	u_i	w_i
2	0, 1	1/2, 1/2
3	0, 1/2, 1	1/4, 1/2, 1/4
4	0, 1/3, 2/3, 1	1/6, 1/3, 1/3, 1/6
5	0, 1/4, 1/2, 3/4, 1	1/8, 1/4, 1/4, 1/4, 1/8

Table 2.1: Weights and nodes associated with trapezoid rule.

2.2   ROMBERG INTEGRATION RULES

The trapezoid rule has the interesting property that the error associated with an estimate obtained from an analysis on N intervals can be expressed entirely in even powers of N [3]. This makes it possible to employ a form of extrapolation attributed to Richardson [3,4] to systematically improve the estimates. As an illustration, consider the approximations $I_{n,1}$ and $I_{n+1,1}$ obtained above. For a smooth integrand, the leading order error drops by a factor of 4, which suggests that we can express the estimates of the integrals as

$$I \cong I_{n,1} + \frac{K}{N^2} \tag{2.11}$$

$$I \cong I_{n+1,1} + \frac{K}{4N^2} \tag{2.12}$$

where K is a constant associated with the error. An improved approximation is possible by solving these equations simultaneously, to determine K and remove the error term from (2.12). The result

$$I \cong I_{n+1,2} = \frac{4}{3}I_{n+1,1} - \frac{1}{3}I_{n,1} \tag{2.13}$$

should be more accurate than (2.11) or (2.12) for smooth integrands. As integral estimates $\{I_{n+1,1}\}$ are generated by the successive application of trapezoid rule, following (2.9), the extrapolation procedure in (2.13) can be immediately applied to generate a sequence $\{I_{n+1,2}\}$. Furthermore, once at least two approximations in this new sequence have been generated from (2.13), the theoretical error analysis implies that the error decreases according

$$I \cong I_{n,2} + \frac{L}{N^4} \tag{2.14}$$

$$I \cong I_{n+1,2} + \frac{L}{16N^4} . \tag{2.15}$$

Equations (2.14) and (2.15) can be solved to produce the improved estimate

$$I \cong I_{n+1,3} = \frac{16}{15}I_{n+1,2} - \frac{1}{15}I_{n,2} . \tag{2.16}$$

If this process of extrapolation is carried to its limit, the result is known as *Romberg quadrature*. A simple subroutine for Romberg integration is provided in Appendix C of [5]. Table 2.2 lists nodes and weights arising from Romberg quadrature for 3 and 5 function samples. These rules are identical to the closed Newton-Cotes rules (3-point Simpson rule and 5-point Boole rule) [4]. In practice, because of the systematic nature of Romberg quadrature and the availability of an error estimate, it has generally superceded the use of Newton-Cotes rules in practice.

Table 2.2: Weights and nodes associated with the Romberg rule.		
nodes	u_i	w_i
3	0, 1/2, 1	1/6, 2/3, 1/6
5	0, 1/4, 1/2, 3/4, 1	7/90, 32/90, 12/90, 32/90, 7/90

The preceding examples involve closed integration rules. Open rules of the Newton-Cotes type are also possible and can be found in the literature [4]. As with the closed rules, these are obtained by interpolating a linear, quadratic, etc., behavior between function samples in order to construct the rule. Table 2.3 lists nodes and weights for several open rules. Note that some of the weights are negative, which is usually not desired due to the possibility of round-off error being exacerbated by the cancellation of large numbers.

Table 2.3: Weights and nodes associated for open Newton-Cotes rules.		
nodes	u_i	w_i
2	1/3, 2/3	1/2, 1/2
3	1/4, 1/2, 3/4	2/3, –1/3, 2/3
4	1/5, 2/5, 3/5, 4/5	11/24, 1/24, 1/24, 11/24
5	1/6, 1/3, 1/2, 2/3, 5/6	33/60, –42/60, 78/60, –42/60, 33/60

2.3   GAUSS-LEGENDRE QUADRATURE RULES

Gauss observed that if the quadrature node locations were not constrained to be equally spaced, the accuracy of the resulting rule for smooth functions can be superior to a Newton-Cotes rule with the same number of nodes. In other words, rather than prescribe the node locations in advance, both the nodes and weights can be selected to optimize the accuracy of the integral estimate.

As an example, consider the general cubic polynomial

$$f(u) = a + bu + cu^2 + du^3 . \tag{2.17}$$

Since there are four degrees of freedom in (2.17), it is reasonable that an integration rule can be found with two node values and two weights. It also happens that left-to-right symmetry can be imposed on both the node locations and the weights. The integral is, therefore, to be approximated by

$$\int_0^1 f(u)du = w_1 f(u_1) + w_1 f(1 - u_1).$$

(2.18)

The nodes and weights of such a rule can be determined, at least in principle, by substituting (2.17) into (2.18) and solving the simultaneous nonlinear equations

$$\int_0^1 adu = a = w_1 a + w_1 a$$

(2.19)

$$\int_0^1 budu = \frac{b}{2} = w_1 b u_1 + w_1 b (1 - u_1)$$

(2.20)

$$\int_0^1 cu^2 du = \frac{c}{3} = w_1 c(u_1)^2 + w_1 c(1 - u_1)^2$$

(2.21)

$$\int_0^1 du^3 du = \frac{d}{4} = w_1 d(u_1)^3 + w_1 d(1 - u_1)^3.$$

(2.22)

These equations are satisfied by the values

$$w_1 = \frac{1}{2}, \ u_1 = \frac{1}{2} - \frac{1}{2\sqrt{3}}$$

(2.23)

$$w_2 = \frac{1}{2}, \ u_2 = 1 - u_1 = \frac{1}{2} + \frac{1}{2\sqrt{3}}.$$

(2.24)

The result in (2.23) and (2.24) is known as the 2-point Gauss-Legendre rule. In a similar manner, a 3-point rule can be developed to exactly integrate a polynomial of degree 5, while a 4-point rule can be found that exactly integrates polynomials up to degree 7. Table 2.4 gives nodes and weights for Gauss-Legendre rules for 2-5 nodes on the unit interval.

The solution of nonlinear equations to obtain quadrature rules is a "brute force" approach. In practice, a more elegant approach based on orthogonal polynomials is normally used to obtain Gauss-Legendre rules. However, for multidimensional integration rules, or for rules based on families of basis functions that do not satisfy appropriate orthogonality conditions, the brute-force approach may be the only available means. A general procedure for solving nonlinear systems of the form of (2.19)–(2.22) to obtain quadrature nodes and weights will be considered in Chapter 5, for developing special quadrature rules that can integrate singular functions.

2.4   GAUSS-LOBATTO QUADRATURE

While Gauss-Legendre rules are of the open variety, alternative forms of Gaussian quadrature such as the Gauss-Radau or Gauss-Lobatto rules place nodes at interval endpoints [6]. Table 2.5 presents weights and notes for Gauss-Lobatto rules.

nodes	u_i	w_i
2	0.21132487	0.5
	0.78867514	0.5
3	0.11270167	0.27777778
	0.5	0.44444444
	0.88729833	0.27777778
4	0.06943185	0.17392743
	0.33000948	0.32607258
	0.66999052	0.32607258
	0.93056816	0.17392743
5	0.04691008	0.11846344
	0.23076534	0.23931434
	.5	0.28444444
	0.76923466	0.23931434
	0.95308992	0.11846344

Table 2.4: Weights and nodes associated with the Gauss-Legendre rule.

2.5   RELATIVE PERFORMANCE OF QUADRATURE RULES

To illustrate the relative accuracy of the various quadrature rules discussed above, we consider their application to an integral associated with the two-dimensional magnetic field integral equation (MFIE), which will be studied in more detail in Chapter 3. The 2D MFIE, for the transverse-magnetic-to-z (TM) polarization, is an equation of the form

$$H_t^{inc}(t) = \frac{J_z(t)}{2} + \int_\Gamma J_z(t') \, C(t,t') dt' \tag{2.25}$$

where H_t^{inc} represents the known incident magnetic field, J_z is the unknown surface current density, t is a parametric variable with units of length along the contour Γ of the 2D structure, and the kernel is

$$C(t,t') = \frac{jk}{4} H_1^{(2)}(kR) \left\{ \frac{[y(t) - y(t')]}{R} \sin \Omega(t) + \frac{[x(t) - x(t')]}{R} \cos \Omega(t) \right\} \tag{2.26}$$

where $x(t)$ and $y(t)$ denote a location on the surface of the structure, $\Omega(t)$ is an angle describing the orientation of the normal vector to the surface, $H_1^{(2)}(kR)$ denotes a Hankel function of the second kind, and

$$R = \sqrt{[x(t) - x(t')]^2 + [y(t) - y(t')]^2} \, . \tag{2.27}$$

Figure 2.1 shows a plot of (2.26) for the special case of a circular domain of radius 0.5 wavelengths, as a function of t' in wavelengths, for an observer located at $t = 0$. Although the

Table 2.5: Weights and nodes associated with the Gauss-Lobatto rule.

nodes	u_i	w_i
3	0.0	0.16666667
	0.5	0.66666667
	1.0	0.16666667
4	0.	0.08333333
	0.27639320	0.41666667
	0.72360680	0.41666667
	1.0	0.08333333
5	0.0	0.05
	0.17267316	0.27222222
	0.5	0.35555555
	0.82732684	0.27222222
	1.0	0.05
6	0.0	0.03333333
	0.11747234	0.18923748
	0.35738424	0.27742919
	0.64261576	0.27742919
	0.88252766	0.18923748
	1.0	0.03333333

function is bounded, a derivative discontinuity in the graph at $t' = 0$ is apparent in the real part of the integrand. The horizontal axis in this plot represents the distance between the observer and the source of the magnetic field along the circular arc; the discontinuity occurs when the source and observer locations coincide.

As a means of evaluating the various quadrature rules introduced in preceding sections, their performance is compared when integrating (2.26) over two different domains of integration, both arising from the circular domain of radius 0.5 wavelengths. (This computation is similar to computing the magnetic field produced by a constant current density.) The first involves an interval from 30 to 60 degrees away from the observer along the circle. The position can be expressed in terms of an equivalent angle given in radians by $\phi = t'/a$. Table 2.6 presents the results. A reference value is obtained by using higher order rules. It is apparent that for these results, which are shifted from the vicinity of the derivative discontinuity, the closed Romberg integration substantially outperforms trapezoid rule and slightly outperforms the open Newton-Cotes rules from Table 2.3. However, the Gauss-Legendre rules yield far better accuracy than any of the others. While all the rules produce estimates of different accuracy between the real and imaginary parts, those differences are slight.

Table 2.7 shows analogous results for a domain from 0 degrees to 30 degrees, which includes the lower limit where the derivative discontinuity in the real part occurs. As compared to the accuracy

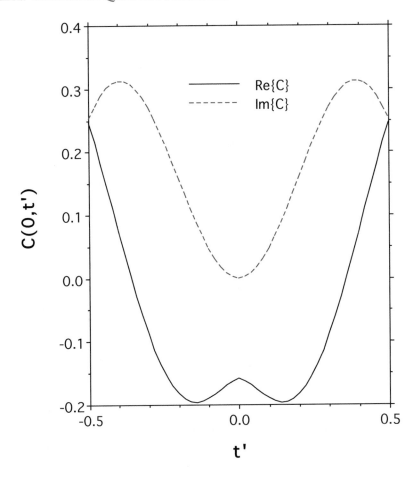

Figure 2.1: Plots of a portion of (2.26) for a circular domain of radius 0.5 wavelength, with $t = 0$.

exhibited in Table 2.6, the trapezoid rule yields similar performance, while the Romberg and open Newton-Cotes rules clearly yield a less accurate result for the real part of the integral. While the Gauss-Lobatto and Gauss-Legendre rules exhibit similar accuracy in Table 2.7 for the imaginary part of the integrals as they did in Table 2.6, their accuracy for the real part in Table 2.7 is worse by as much as 4 orders of magnitude. Clearly, the presence of the derivative discontinuity has a dramatic effect on the integration accuracy of the Gauss-Legendre rule. Despite this, the Gauss-Legendre rule remains the most accurate of the rules considered for 3 or more quadrature nodes.

The preceding comparisons suggest that, for relatively smooth functions, the Gauss-Legendre rules with 3 or more nodes produce far more accurate integral estimates than the Newton-Cotes or Romberg rules. Thus, in Chapter 3, we will focus on the use of Gauss-Legendre rules when initially

Table 2.6: Performance of quadrature rules when integrating $C(0, t')$ for a circular domain of radius 0.5 wavelengths, over a range of angles between 30 and 60 degrees.

	Re{C}	Error	Im{C}	Error
		Trapezoid rule:		
2	0.01928	12.0 %	0.05972	19.4 %
3	0.01771	2.9	0.07072	4.7
4	0.01744	1.3	0.07265	2.1
5	0.01734	0.7	0.07332	1.2
		Romberg:		
3	0.0171902	0.14 %	0.0743810	0.28 %
5	0.0172148	0.0016	0.0741701	0.0017
		Open Newton-Cotes:		
2	0.0165104	4.1 %	0.0791110	6.7 %
3	0.0172363	0.12	0.0739856	0.25
4	0.0172300	0.09	0.0740431	0.17
5	0.0172157	0.0033	0.0741741	0.0036
		Gauss-Lobatto:		
3	0.01719022	0.14 %	0.07438103	0.28 %
4	0.01721474	0.0021	0.07416977	0.0022
5	0.01721510	2.1×10^{-5}	0.07417143	1.1×10^{-5}
		Gauss-Legendre:		
2	0.01723143838	0.10 %	0.07403055110	0.19 %
3	0.01721536489	0.0016	0.07417267085	0.0017
4	0.01721509442	1.7×10^{-5}	0.07417142054	9×10^{-6}
5	0.01721509741	1.8×10^{-7}	0.07417142697	3×10^{-8}
		Reference:		
	0.017215097379		0.074171426946	

implementing the Nyström procedure. The results also suggest that the dramatic improvement in accuracy obtained with Gauss-Legendre rules as the number of nodes is increased for smooth functions (Table 2.6 and the imaginary part of the integral in Table 2.7) is lost if the integrand in question exhibits non-analytical behavior, even for something as mild as a derivative discontinuity at the end of the integration interval. This observation, and the relatively poor performance of the conventional Nyström method when used to discretize the MFIE, motivate the development of the locally-corrected Nyström method (Chapter 4).

Table 2.7: Performance of quadrature rules when integrating $C(0, t')$ for a circular domain of radius 0.5 wavelengths, over a range of angles between 0 and 30 degrees.

	Re{C}	Error	Im{C}	Error
Trapezoid rule:				
2	−0.03844	17.5 %	0.03046	28.2 %
3	−0.04487	3.7	0.02534	6.7
4	−0.04588	1.6	0.02445	2.9
5	−0.04621	0.9	0.02414	1.6
Romberg:				
3	−0.047013	0.85 %	0.023627	0.54 %
5	−0.046632	0.035	0.023755	0.003
Open Newton-Cotes:				
2	−0.049593	6.4 %	0.021445	9.7 %
3	−0.046299	0.68	0.023867	0.47
4	−0.046389	0.49	0.023832	0.33
5	−0.046597	0.040	0.023753	0.006
Gauss-Lobatto:				
3	−0.047013	0.85 %	0.023627311	0.53 %
4	−0.046635	0.042	0.023755542	0.0039
5	−0.046620	0.0092	0.023754616	1.7×10^{-5}
Gauss-Legendre:				
2	−0.0463693	0.53 %	0.023840127881	0.36 %
3	−0.0466043	0.024	0.023753926916	0.0029
4	−0.0466130	0.0057	0.023754623581	1.4×10^{-5}
5	−0.0466149	0.0016	0.023754620345	4×10^{-8}
Reference:				
	−0.046615664116		0.023754620355	

REFERENCES

[1] E. Maor, *e: The Story of a Number*. Princeton: Princeton University Press, 1994, p. 58.

[2] D. Kahaner, C. Moler, and S. Nash, *Numerical Methods and Software*. Englewood Cliffs: Prentice-Hall, 1989.

[3] W. Cheney and D. Kincaid, *Numerical Mathematics and Computing*. Monterey: Brooks/Cole Publishing, 1980.

[4] K. E. Atkinson, *An Introduction to Numerical Analysis*. New York: Wiley, 1989.

[5] A. F. Peterson, S. L. Ray, and R. Mittra, *Computational Methods for Electromagnetics*. New York: IEEE Press, 1998.

[6] F. B. Hildebrand, *Introduction to Numerical Analysis*. New York: Dover, 1987.

CHAPTER 3

The Classical Nyström Method

In this chapter, the classical Nyström method is illustrated using the two-dimensional (2D) magnetic field integral equation (MFIE) for the currents on perfectly conducting structures. For circular cylinder targets, a flat-cell faceted discretization is compared to using the actual circular surface. While we primarily employ open Gauss-Legendre quadrature, we also illustrate the use of closed trapezoid/Romberg quadrature rules.

3.1 THE MAGNETIC FIELD INTEGRAL EQUATION

Derivations and descriptions of the MFIE are provided in [1–3]. In the 2D case, there are two polarizations of interest, the transverse-magnetic-to-z (TM) and the transverse-electric-to-z (TE). The equation for the TM polarization can be expressed in the form

$$H_t^{\text{inc}}(t) = \frac{J_z(t)}{2} + \int_\Gamma J_z(t')\, C_1(t, t') dt' \tag{3.1}$$

where H_t^{inc} represents the known incident magnetic field (defined in the absence of the target), J_z is the unknown surface current density, t is a parametric variable with units of length along the contour Γ of the 2D structure, and the kernel is

$$C_1(t, t') = \begin{cases} -\frac{\kappa(t)}{4\pi} & \lim t' \to t \\ \frac{jk}{4} H_1^{(2)}(kR) \left\{ \frac{[y(t)-y(t')]}{R} \sin \Omega(t) + \frac{[x(t)-x(t')]}{R} \cos \Omega(t) \right\} & \text{otherwise} \end{cases} . \tag{3.2}$$

In (3.2), κ denotes the curvature (the reciprocal of the radius of curvature)

$$\kappa(t) = \frac{1}{\rho(t)} \tag{3.3}$$

at the point $x(t)$, $y(t)$ on the cylinder contour, $H_1^{(2)}(\bullet)$ is the first-order Hankel function of the second kind, k is the medium wavenumber, and

$$R = \sqrt{[x(t) - x(t')]^2 + [y(t) - y(t')]^2} . \tag{3.4}$$

The outward normal vector along the contour Γ is defined in terms of an angle Ω (the angle between the x-axis and the normal) by

$$\hat{n}(t) = \hat{x} \cos \Omega(t) + \hat{y} \sin \Omega(t) . \tag{3.5}$$

The transverse tangent vector is similarly obtained as

$$\hat{t}(t) = -\hat{x} \sin \Omega(t) + \hat{y} \cos \Omega(t) . \tag{3.6}$$

For our purposes, the incident field is a uniform plane wave propagating into a polar angle θ^{inc} in the x-y plane, with a transverse component that can be expressed at any point on the contour as

$$H_t^{\text{inc}}(t) = -\cos\{\theta^{\text{inc}} - \Omega(t)\}e^{-jk(x \cos \theta^{\text{inc}} + y \sin \theta^{\text{inc}})} \tag{3.7}$$

where $x = x(t)$, etc. The MFIE in (3.1) is expressed in a form that restricts the contour Γ to be smooth; in Chapter 6, we will generalize the approach to contours with corners.

For the TE polarization, the MFIE for smooth contours is

$$H_z^{\text{inc}}(t) = -\frac{J_t(t)}{2} - \int_\Gamma J_t(t')C_2(t, t')dt' \tag{3.8}$$

where H_z^{inc} is the known incident field defined in the absence of the target, J_t is the unknown surface current density, and the kernel is

$$C_2(t, t') = \begin{cases} \frac{\kappa(t)}{4\pi} & \lim t' \to t \\ \frac{jk}{4} H_1^{(2)}(kR) \left\{ \frac{[y(t)-y(t')]}{R} \sin \Omega(t') + \frac{[x(t)-x(t')]}{R} \cos \Omega(t') \right\} & \text{otherwise} \end{cases} . \tag{3.9}$$

The incident field may be written as

$$H_z^{\text{inc}}(t) = e^{-jk(x \cos \theta^{\text{inc}} + y \sin \theta^{\text{inc}})} . \tag{3.10}$$

An inspection of Equations (3.1) and (3.8) indicates that their integrands are bounded; thus the original Nyström method [4,5] may be used to generate numerical solutions of these equations.

3.2   FLAT-FACETED DISCRETIZATION

To simplify an initial presentation of the Nyström method, we employ the common (but, as we will soon see, unfortunate) practice of representing the scatterer contour by flat facets. Suppose that the 2D scatterer contour is divided into flat cells that are small with respect to the wavelength λ. The essence of the Nyström method is that within each cell, the integral within (3.1) is approximated by a quadrature summation

$$\int_{\text{cell } n} J_z(t') C_1(t, t')dt' \cong \sum_{i=1}^q w_{ni} J_z(t_{ni}) C_1(t, t_{ni}) \tag{3.11}$$

where $\{w_{ni}, t_{ni}\}$ denote the weights and nodes of a suitable q-point quadrature rule used for cell n. For illustration, conventional Gauss-Legendre quadrature rules are used[1]. The samples $J_{ni} = J_z(t_{ni})$ are

---

[1] As discussed in Chapter 2, the derivative discontinuity in the MFIE kernel results in limited accuracy for high-order Gauss-Legendre rules.

the unknowns to be determined. (Since these samples represent the sources of the electromagnetic fields, we sometimes refer to the specific cell containing the samples of interest as the *source* cell.) If there are N cells around the contour, and a q-point rule is used uniformly, there are Nq unknown samples to be determined. These unknowns are determined by imposing Equation (3.1) in Nq linearly independent ways. Since the integrand in (3.1) is bounded, the equation can be enforced at each the nodes of the quadrature rule in each cell. Each of the nodes where the integral equation is imposed is known as an *observer* location. The result is an Nq by Nq system of equations:

$$
\begin{bmatrix} H_{11}^{\text{inc}} \\ \vdots \\ H_{1q}^{\text{inc}} \\ H_{21}^{\text{inc}} \\ \vdots \\ H_{Nq}^{\text{inc}} \end{bmatrix} = \begin{bmatrix} \left(\frac{1}{2}+Y_{11,11}\right) & \cdots & Y_{11,1q} & Y_{11,21} & \cdots & Y_{11,Nq} \\ \vdots & \ddots & \vdots & \vdots & & \vdots \\ Y_{1q,11} & \cdots & \left(\frac{1}{2}+Y_{1q,1q}\right) & Y_{1q,21} & \cdots & Y_{1q,Nq} \\ Y_{21,11} & \cdots & Y_{21,1q} & \left(\frac{1}{2}+Y_{21,21}\right) & \cdots & Y_{21,Nq} \\ \vdots & & \vdots & \vdots & \ddots & \vdots \\ Y_{Nq,11} & \cdots & Y_{Nq,1q} & Y_{Nq,21} & \cdots & \left(\frac{1}{2}+Y_{Nq,Nq}\right) \end{bmatrix} \begin{bmatrix} J_{11} \\ \vdots \\ J_{1q} \\ J_{21} \\ \vdots \\ J_{Nq} \end{bmatrix} . \tag{3.12}
$$

If a flat-cell faceted model is used,

$$
Y_{mj,mi} = 0 \tag{3.13}
$$

whenever indices representing the source and observer are located in the same cell, as a consequence of the property that a planar current density produces no tangential magnetic field in that same plane. The matrix entries for $m \neq n$ (whenever the source and observer cells are different) are given by

$$
Y_{mj,ni} = \frac{jk}{4} w_i \left[\sin \Omega_m \frac{y_{mj} - y_{ni}}{R_{mj,ni}} + \cos \Omega_m \frac{x_{mj} - x_{ni}}{R_{mj,ni}} \right] H_1^{(2)}(k R_{mj,ni}) . \tag{3.14}
$$

The incident field is sampled at the observer locations, yielding

$$
H_{mj}^{\text{inc}} = -\cos\{\theta^{\text{inc}} - \Omega_m\} e^{-jk(x_{mj}\cos\theta^{\text{inc}} + y_{mj}\sin\theta^{\text{inc}})} . \tag{3.15}
$$

In (3.14) and (3.15), coordinates are denoted $x_{mj} = x(t_{mj})$, etc.,

$$
R_{mj,ni} = \sqrt{[x_{mj} - x_{ni}]^2 + [y_{mj} - y_{ni}]^2} . \tag{3.16}
$$

and Ω_m is used to denote the orientation of cell m. The result in (3.13) is exact for flat cells and simplifies the formulation so that all the diagonal entries of (3.12) have numerical values of 0.5.

As an example, consider a circular conducting cylinder of radius a. Flat-cell models are easily constructed and may be described by a minimum amount of information, such as the central coordinates (x_n, y_n) of each cell, the transverse length d_n of the cell, and the orientation Ω_n of the cell. (It is usually advantageous to scale the cells so that the circumference of the model equals that of the desired cylinder, and that approach was used here.) Using these parameters, and the quadrature rule $\{w_{ni}, t_{ni}\}$, the matrix entries specified in (3.13)–(3.15) are easily computed. The user controls the number of cells in the model and the order q of the quadrature rule. Although the order of the quadrature may be different in different cells, this illustration uses a uniform order for simplicity. We will use the number of nodes q in a quadrature rule to denote the "order" of the rule; note that Gaussian Legendre rules with q nodes can exactly integrate polynomials of degree $2q - 1$.

Consider a circular cylinder with $ka = 3.5$, where a is the radius. Table 3.1 shows the current density at the specular point on the cylinder, produced by the Nyström formulation with $q = 3$ and models with 10, 20, 40, and 80 equal-size cells. Note that these models produce a total number of unknowns equal to 30, 60, 120, and 240, respectively. The models are adjusted so that the center node of the quadrature rule in one cell coincides with the specular point in each case. The relative error

$$\text{error} = \frac{|J_{\text{exact}} - J_{\text{numerical}}|}{|J_{\text{exact}}|} \tag{3.17}$$

is also shown for each result, based on the exact solution for the circular cylinder geometry [6]. We will comment on the level of error below.

Table 3.1: Nyström MFIE results for the TM current density at the specular point ($\phi = 180$) on a circular cylinder with $ka = 3.5$ induced by a uniform plane wave with $\theta^{\text{inc}} = 0$. Results are obtained with a $q = 3$ Gauss-Legendre quadrature rule and flat-cell models.

Number of cells	magnitude	phase	Relative error
10	1.8538	−172.68	0.1443
20	1.9445	−168.16	6.490×10^{-2}
40	2.0023	−166.96	3.171×10^{-2}
80	2.0336	−166.65	1.586×10^{-2}
exact [6]	2.066168	−166.5489	

The Nyström discretization of the MFIE for the TE polarization may be obtained in a similar manner, yielding a system of equations of the form

$$\begin{bmatrix} H_{11}^{\text{inc}} \\ \vdots \\ H_{1q}^{\text{inc}} \\ H_{21}^{\text{inc}} \\ \vdots \\ H_{Nq}^{\text{inc}} \end{bmatrix} = \begin{bmatrix} \left(-\frac{1}{2} + Y_{11,11}\right) & \cdots & Y_{11,1q} & Y_{11,21} & \cdots & Y_{11,Nq} \\ \vdots & \ddots & \vdots & \vdots & & \vdots \\ Y_{1q,11} & \cdots & \left(-\frac{1}{2} + Y_{1q,1q}\right) & Y_{1q,21} & \cdots & Y_{1q,Nq} \\ Y_{21,11} & \cdots & Y_{21,1q} & \left(-\frac{1}{2} + Y_{21,21}\right) & \cdots & Y_{21,Nq} \\ \vdots & & \vdots & \vdots & \ddots & \vdots \\ Y_{Nq,11} & \cdots & Y_{Nq,1q} & Y_{Nq,21} & \cdots & \left(-\frac{1}{2} + Y_{Nq,Nq}\right) \end{bmatrix} \begin{bmatrix} J_{11} \\ \vdots \\ J_{1q} \\ J_{21} \\ \vdots \\ J_{Nq} \end{bmatrix} \tag{3.18}$$

where (again as a consequence of flat cells)

$$Y_{mj,mi} = 0 \tag{3.19}$$

when the source and observer share the same cell,

$$Y_{mj,ni} = -\frac{jk}{4} w_i \left[\sin \Omega_n \frac{y_{mj} - y_{ni}}{R_{mj,ni}} + \cos \Omega_n \frac{x_{mj} - x_{ni}}{R_{mj,ni}} \right] H_1^{(2)}(k R_{mj,ni}), \; m \neq n \tag{3.20}$$

$$H_{mj}^{\text{inc}}(t) = e^{-jk(x_{mj} \cos \theta^{\text{inc}} + y_{mj} \sin \theta^{\text{inc}})} \tag{3.21}$$

and where $J_{ni} = J_t(t_{ni})$ now denotes samples of the transverse current density. Observe that in (3.20) the orientation angles are a function of the source cell index n, compared with those of (3.14) that were a function of the observer cell index m.

For a circular cylinder of dimension ka = 3.5, Table 3.2 shows results for the current density produced at the specular point in response to a TE plane wave illumination, obtained with a Gauss-Legendre quadrature rule of order q = 3 and various equal-cell-size models. As observed for the TM polarization, the relative error is decreasing as the number of cells in the model increases. It is noted that the TE error appears to decrease faster than the TM error reported in Table 3.1.

Table 3.2: Nyström MFIE results for the TE current density at the specular point ($\phi = 180$) on a circular cylinder with ka = 3.5 induced by a uniform plane wave with $\theta^{\text{inc}} = 0$. Results are obtained with q = 3 Gauss-Legendre quadrature rules and flat-cell models.

Number of cells	magnitude	phase	Relative error
10	1.9319	20.95	0.1144
20	1.9125	26.06	2.419×10^{-2}
40	1.9110	27.18	4.655×10^{-3}
80	1.9113	27.42	5.078×10^{-4}
exact [6]	1.911749	27.4458	

There are several aspects of the Nyström procedure worth mentioning. First, the ultimate goal is usually to obtain a good approximation of the current density on the structure. The accuracy of the current density is primarily determined by its representation. In this case, the representation of the current is *implicit* since we are not directly expressing it in terms of a set of explicit basis functions. In the Nyström approach, the representation of the current is indirectly determined by the choice of quadrature rule. Since Gauss-Legendre quadrature is designed to exactly integrate polynomials up to a certain degree, one could postulate that the current representation is essentially whatever the quadrature rule can integrate — say, polynomials up to that degree.

Our implementation uses a subsectional model of the computational domain. Since an open quadrature rule is separately imposed over each cell, there is no explicit enforcement of the continuity of current density between cells. Thus, the representation is somewhat equivalent to an independent polynomial expansion within each cell.

Therefore, there are two ways to improve the accuracy of the Nyström procedure: increase the order q of the quadrature rule or increase the resolution of the model by using smaller cells. As observed in Chapter 2, the improvement in accuracy in moving from a 3-point Gauss-Legendre rule to a 5-point rule, for sufficiently smooth functions, is often considerable. The effect of refining the models by decreasing the cell sizes has already been illustrated in Tables 3.1 and 3.2 where a steady but modest improvement is noted.

The relative accuracy of numerical results is often assessed by plotting some measure of the error as a function of the number of unknowns. This is straightforward for the circular cylinder examples since an exact reference solution is available [6]. For instance, Figure 3.1 shows plots of the error in the current density as a function of the total number of unknowns Nq, for values of q equal to 1, 3, and 5. In this case, the error in the current density is evaluated at all the samples around the cylinder circumference, and the 2-norm error

$$
\text{error} = \frac{\sqrt{\dfrac{1}{Nq} \displaystyle\sum_{n=1}^{N} \sum_{i=1}^{q} |J_{\text{exact}}(t_{ni}) - J_{ni}|^2}}{|J_{\text{exact}}|_{\max}}
\tag{3.22}
$$

is computed. Figure 3.1 shows a log-log plot of the relative error for the TM case. It is noteworthy that the accuracy obtained for $q=1$ for a given number of unknowns is superior to that obtained for $q=3$ and $q=5$, which runs counter to our postulate that improved accuracy should be obtained with a quadrature rule of higher order. In fact, the relative error numbers computed in Table 3.1 for the TM case indicate that the accuracy of the current at the specular point is only improving at an approximate rate of $O(h)$, where h is the cell size in wavelengths. The curves in Figure 3.1 all reflect an $O(h)$ slope, or equivalently an $O(M^{-1})$ rate, where M is the total number of unknowns. These results suggest that we are not realizing the benefit of higher order representations.

Figure 3.2 shows similar plots for the 2-norm error in the TE currents, for q equal to 1, 3, and 5. For $q=1$, the TE and TM results exhibit similar accuracy levels as a function of $M = Nq$. However, for $q=3$, the TE results are more accurate than the TM results and appear to follow an $O(M^{-2})$ slope. In the TM case, the $q=5$ results exhibit a slightly larger error for the same models than the $q=1$ and $q=3$ results, while in the TE case the $q=5$ and $q=3$ results are substantially more accurate than the $q=1$ results but exhibit similar error levels for a given model. Since the $q=5$ results involve the use of five times as many unknowns as $q=1$, obtaining similar accuracy for the same model implies a substantial decrease in the accuracy per unknown! Obviously, something is not working as expected.

Although the equations in (3.1) and (3.8) appear to be similar, there is one fundamental difference in the behavior of the current density for the two polarizations that might be the source of the different error levels. The current density in the TM case is known to be unbounded in the

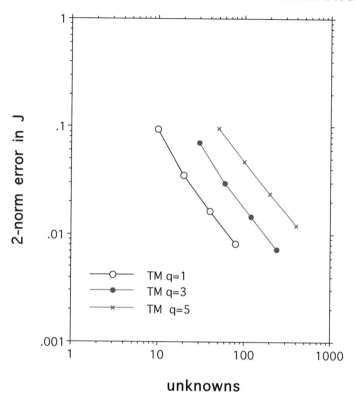

Figure 3.1: The error in the TM surface current density produced by the Nyström method using flat-cell models for a circular cylinder with $ka = 3.5$ and quadrature rules of order $q=1$, $q=3$, and $q=5$, as defined by (3.22).

vicinity of sharp corners, while that of the TE polarization remains finite (but in such a way that the charge density associated with the current becomes unbounded at corners in the TE case). By modeling the circular contour with flat facets, we introduced corners into the actual geometry being analyzed and, therefore, introduced fictitious corner singularities in the current density.

Figure 3.3 shows a portion of a plot of the current density values obtained from the Nyström approach for $q=5$, for a 40-cell flat-faceted model, compared with the exact solution for a circular cylinder with $ka = 3.5$. The plot clearly shows that the current samples near the cell edges exhibit magnitude peaks, suggesting that the numerical procedure is attempting to model the presence of an edge between adjacent cells. The singularities associated with each corner introduced by the faceted model negate the accuracy associated with the quadrature rule and, consequently, reduce the accuracy of the underlying representation of the current for $q=3$ and $q=5$. For the TM case, the best accuracy is obtained with $q=1$ (which, for a given total number of unknowns, involves the best faceted model

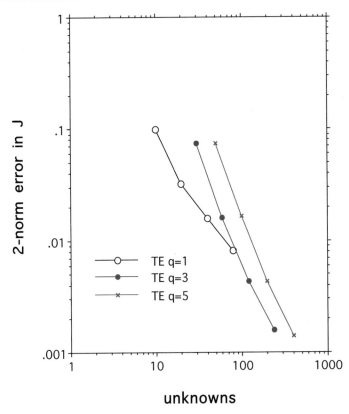

Figure 3.2: The error in the TE surface current density produced by the Nyström method using flat-cell models for a circular cylinder with $ka = 3.5$ and quadrature rules of order $q=1$, $q=3$, and $q=5$, as defined by (3.22).

of the cylinder under consideration). In order to investigate the impact of the fictitious edges, we next consider an improved model of the cylinder where cells actually conform to the circular contour.

3.3   DISCRETIZATION USING EXACT MODELS OF A CIRCULAR CYLINDER

There are several ways that the representation of a circle can be incorporated into the Nyström discretization procedure. One approach is to transform the integral equations directly into cylindrical coordinates and specialize each to a circular geometry. An alternative approach, that is somewhat

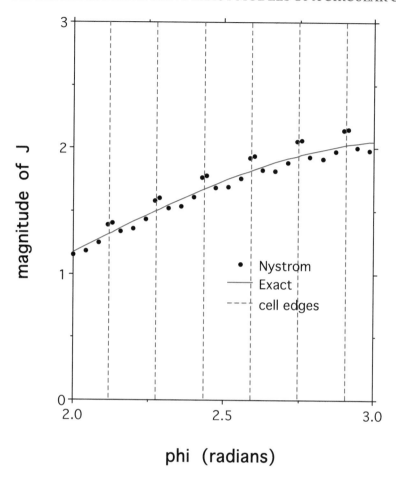

Figure 3.3: The TM surface current density produced by the Nyström method along a portion of the cylinder surface, for a circular cylinder with $ka = 3.5$. A model employing 40 flat cells is used with a quadrature rule of order $q=5$.

more general, is to describe the cylinder geometry parametrically by coordinates

$$x(t) = a \cos\left(\frac{t}{a}\right) \tag{3.23}$$

$$y(t) = a \sin\left(\frac{t}{a}\right) \tag{3.24}$$

where t again denotes a variable coinciding with the arc-length (observe that t/a is the conventional cylindrical angle ϕ). We proceed to follow the Nyström approach as described in Section 3.2, to produce an equation of the form of (3.12) for the TM case. The entries of the matrix equation are identical in form to those used previously. For the source and observer in the same cell ($m = n$),

$$Y_{mj,mj} = -\frac{\kappa(t_{mj})}{4\pi} w_j = \frac{-1}{4\pi a} w_j \tag{3.25}$$

$$Y_{mj,mi} = \frac{jk}{4} w_i \left[\sin \Omega_m \frac{y_{mj} - y_{mi}}{R_{mj,mi}} + \cos \Omega_m \frac{x_{mj} - x_{mi}}{R_{mj,mi}} \right] H_1^{(2)}(k R_{mj,mi}), \quad i \neq j \tag{3.26}$$

and when the source and observer are in different cells ($m \neq n$)

$$Y_{mj,ni} = \frac{jk}{4} w_i \left[\sin \Omega_m \frac{y_{mj} - y_{ni}}{R_{mj,ni}} + \cos \Omega_m \frac{x_{mj} - x_{ni}}{R_{mj,ni}} \right] H_1^{(2)}(k R_{mj,ni}), \quad m \neq n \tag{3.27}$$

where, if t_{ni} denotes quadrature node i in cell n, the coordinates are defined in a similar manner, e.g., $x_{ni} = x(t_{ni})$. These coordinates are obtained from (3.23) and (3.24), so they lie on the actual circle. The quadrature weights must be adjusted to give the appropriate arc-length for the range of t used within each cell[2]. The incident field is not changed from that given in (3.15).

For the TE polarization, the procedure yields a system of the form of (3.18), with same-cell entries

$$Y_{mj,mj} = -\frac{1}{4\pi a} w_j \tag{3.28}$$

$$Y_{mj,mi} = -\frac{jk}{4} w_i \left[\sin \Omega_m \frac{y_{mj} - y_{mi}}{R_{mj,mi}} + \cos \Omega_m \frac{x_{mj} - x_{mi}}{R_{mj,mi}} \right] H_1^{(2)}(k R_{mj,mi}), \quad i \neq j . \tag{3.29}$$

When the source and observer are in separate cells ($m \neq n$),

$$Y_{mj,ni} = -\frac{jk}{4} w_i \left[\sin \Omega_n \frac{y_{mj} - y_{ni}}{R_{mj,ni}} + \cos \Omega_n \frac{x_{mj} - x_{ni}}{R_{mj,ni}} \right] H_1^{(2)}(k R_{mj,ni}), \quad m \neq n . \tag{3.30}$$

The TE incident field is not changed from that in (3.21).

For illustration, we again consider a circular cylinder with $ka = 3.5$. Figures 3.4 and 3.5 present the error in the current density for the TM and TE case, respectively, for Gauss-Legendre quadrature rules of order $q=1$, $q=3$, and $q=5$. Models consisting of 5, 10, 20, and 40 cells, with points on the actual circle, are employed. Compared to the flat-cell results in Figures 3.1 and 3.2, we observe a substantial improvement in accuracy in Figures 3.4 and 3.5. For the TM case, the error in the result obtained with approximately 100 unknowns ($Nq \cong 100$) is about three orders of magnitude better with the circular models than with the flat-cell models. The error curves obtained with the circular-cell models are similar for both the TM and TE polarizations. This suggests that eliminating the fictitious corners introduced by the flat-cell models considered previously has improved the accuracy, and in the absence of the edges, the MFIE performs similarly for the TM and TE situations.

---

[2]For quadrature rules defined on a unit interval, an additional Jacobian may be included in the integrands to compensate, and the nodes must be shifted to appropriate locations in the cells.

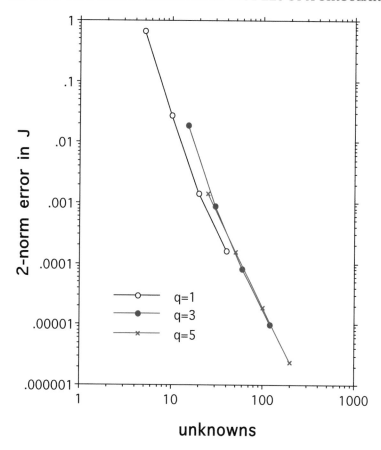

Figure 3.4: The error in the TM surface current density produced by the Nyström method using an exact model of a circular cylinder with $ka = 3.5$ and quadrature rules of order $q=1$, $q=3$, and $q=5$, as defined by (3.22).

However, the $q=3$ and $q=5$ error curves are similar to each other for both the TM and TE polarizations and all decrease at an approximate $O(M^{-3})$ rate as the number of unknowns M is increased. This suggests that quadrature rules beyond $q = 3$ do not improve the accuracy of the numerical results even for an exact model of the circular geometry. This limitation on accuracy appears to be a consequence of the MFIE kernel singularity, as discussed previously.

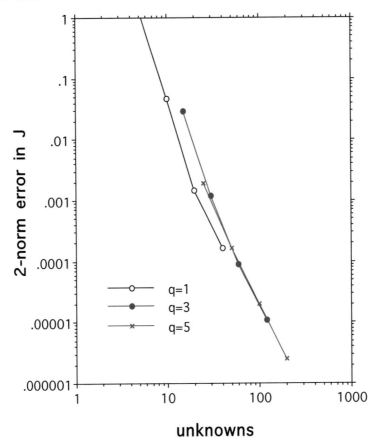

Figure 3.5: The error in the TE surface current density produced by the Nyström method using an exact model of a circular cylinder with $ka = 3.5$ and quadrature rules of order $q=1$, $q=3$, and $q=5$, as defined by (3.22).

3.4   NYSTRÖM DISCRETIZATIONS USING CLOSED QUADRATURE RULES

The previous results were obtained with Nyström discretizations employing Gauss-Legendre quadrature rules, which are open rules (the set of nodes does not include the endpoints of the interval). One consequence of the use of open rules is that cell-to-cell continuity of the current density is *not* explicitly imposed by the representation. For smooth, closed cylinders, a Nyström discretization of the 2D MFIE may also be based on closed quadrature rules, such as the trapezoid rule or the Romberg rule. The use of closed rules implies that one sample of the current will be located at each cell boundary/junction, and shared by both adjoining cells, effectively imposing cell-to-cell current continuity.

The implementation of closed rules differs only slightly from that of open rules. Entries of the system matrix corresponding to a source node located in the interior of a cell are similar to those described earlier in this chapter while a source node located at the boundary between two cells contributes to the quadrature summations for both cells.

The performance of trapezoid rule is illustrated in Figure 3.6 for a circular cylinder with ka = 3.5, modeled with curved cells and illuminated by a TM uniform plane wave. Results are shown

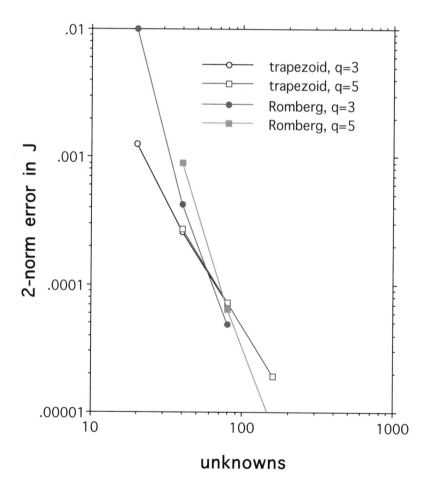

Figure 3.6: The error in the TM surface current density produced on a circular cylinder with ka = 3.5 by the Nyström method using 3-point and 5-point trapezoid rule (open markers), and 3-point and 5-point Romberg integration rules (closed markers). The error is defined by (3.22).

for rules with 3 and 5 points per cell, and for models with 10, 20, and 40 cells. The accuracy of the surface current density improves as the number of cells increases. As might be expected, there is no advantage in the use of "higher-order" trapezoid rules (the underlying representation, of course, is the same for a given distribution of nodes). The error appears to decrease as $O(M^{-2})$ as the number of unknowns M (the total number of nodes along the cylinder contour) is increased.

As a second illustration, we also consider the use of closed Romberg quadrature, with 3-point and 5-point rules. For comparison purposes, the errors arising from these rules are also shown in Figure 3.6 for the same TM circular cylinder with $ka = 3.5$. Compared with trapezoid rule, the Romberg quadrature appears to yield an improved convergence rate of $O(M^{-3})$ as the number of unknowns M is increased. Even so, the Romberg results are slightly less accurate than the Gauss-Legendre results (Figure 3.4) for a given number of unknowns, with an error that decreases at a similar rate.

These two examples suggest that there is no advantage to the use of closed quadrature rules within the Nyström discretization of the MFIE, despite the fact that current continuity is imposed as a by-product of their use. This issue will be explored further in Chapter 4.

3.5   SUMMARY

The conventional Nyström discretization procedure has been described and used to provide numerical solutions for the 2D MFIE representing perfectly conducting targets. The method was initially illustrated using Gauss-Legendre quadrature rules in conjunction with flat-faceted models, which introduce singularities at the fictitious edges and result in a relatively slow rate of convergence. Of interest is the fact that the results for the TM-to-z polarization are significantly worse than those for the TE-to-z polarization as the quadrature order is increased, perhaps due to the different nature of the current singularity at an edge for these polarizations. Exact circular models were also employed, and the resulting Nyström solutions exhibit far greater accuracy compared with the flat cell models, and similar accuracy for TM and TE polarizations. We also considered the implementation of closed quadrature rules (trapezoid and Romberg). Despite the fact that these approaches imposed cell-to-cell continuity of the current density, they provided no improvement in overall accuracy compared to the use of open Legendre rules.

Although the Nyström solutions for circular cylinders improved by several orders of magnitude when exact models replaced the flat-faceted models of the cylinder surface, they still did not converge at continually faster rates as the Gauss-Legendre quadrature rule was refined. This is apparently a result of the derivative discontinuity in the MFIE kernel, which limits the accuracy of the quadrature rules as discussed in Chapter 2. It is noteworthy that the error rate of $O(M^{-3})$ observed with the Gauss-Legendre and Romberg rules is better than that obtained with many alternative low-order MoM procedures, and, therefore, those approaches (at least up to $q = 3$) may prove useful. If additional accuracy is required, the conventional Nyström procedure must be extended in order to treat kernels with derivative discontinuities or stronger singularities. One way of accomplishing this is the "locally corrected" Nyström (LCN) approach, described in Chapter 4.

REFERENCES

[1] A. J. Poggio and E. K. Miller, "Integral equation solutions of three-dimensional scattering problems," in *Computer Techniques for Electromagnetics*, ed. R. Mittra, New York: Hemisphere, 1987 reprint.

[2] A. F. Peterson, S. L. Ray, and R. Mittra, *Computational Methods for Electromagnetics*. New York: IEEE Press, 1998.

[3] K. Warnick, *Numerical Analysis for Electromagnetic Integral Equations*. Boston: Artech House, 2008.

[4] E. J. Nyström, "Über die praktische Auflüsung von Integral-gleichungen mit Anwendungen auf Randwertaufgaben," *Acta Math.*, vol. 54, pp. 185–204, 1930. DOI: 10.1007/BF02547521

[5] K. E. Atkinson, *A Survey of Numerical Methods for the Solution of Fredholm Integral Equations of the Second Kind*. Philadelphia: SIAM, 1976, pp. 88–105.

[6] R. F. Harrington, *Time-harmonic Electromagnetic Fields*. New York: McGraw-Hill, 1961, pp. 232–235.

CHAPTER 4

The Locally-Corrected Nyström Method

In this chapter, the Nyström method is extended in order to better handle singularities in the kernel of the integral operator. The extension, known as the "locally corrected" Nyström (LCN) method, was originally proposed in [1–3]. In contrast to the conventional Nyström method, the LCN enables a high accuracy discretization of integral operators whose kernels have either weak or strong singularities. The LCN will be applied to two-dimensional electromagnetic scattering problems using the electric-field and magnetic-field integral equations.

4.1   THE LOCALLY-CORRECTED NYSTRÖM METHOD

A fundamental difficulty with the classical Nyström method is that it fails if the kernel of the integral operator is infinite at one of the quadrature sample points. Although the MFIE considered in Chapter 3 has a bounded kernel, most of the operators of interest in electromagnetics have kernels with stronger singularities (in fact, most are infinite when the source and observer locations coincide). As demonstrated in Chapter 3, even if the kernel is bounded, the fact that it is not analytic degrades the accuracy of the overall Nyström solution.

One remedy to this situation was proposed in [1–3], based on an approach for enhancing the accuracy of quadrature rules [4]. The original idea was to synthesize a new quadrature rule when the source and observer locations are in close proximity. An essentially equivalent approach is to synthesize a new (bounded and smooth) kernel to be sampled by the original quadrature rule for near-field interactions. In equation form, the original approximation of the integral operator over a cell (using the notation from Chapter 3)

$$\int_{\text{cell } n} J_z(t')\, K(t_{mj}, t')dt' \cong \sum_{i=1}^{q} w_{ni}\, J_z(t_{ni})\, K(t_{mj}, t_{ni}) \tag{4.1}$$

is replaced by

$$\int_{\text{cell } n} J_z(t')\, K(t_{mj}, t')dt' \cong \sum_{i=1}^{q} w_{ni}\, J_z(t_{ni})\, L(t_{mj}, t_{ni}) \tag{4.2}$$

where L is the new "corrected" kernel, which must be synthesized at the necessary samples. One way to accomplish that is to derive L so that the near fields of some set of hypothetical current distributions are correct at the necessary sample points. The hypothetical currents are, in essence, just a set of basis functions $\{B_k(t)\}$. By imposing the condition that

$$\sum_{i=1}^{q} w_{ni} B_k(t_{ni}) \, L(t_{mj}, t_{ni})$$

$$= \sum_{i=1}^{q} B_k(t_{ni}) \left\{ w_{ni} L(t_{mj}, t_{ni}) \right\} \tag{4.3}$$

$$= \int_{\text{cell } n} B_k(t') \, K(t_{mj}, t') dt'$$

for each basis function in the set, and each location t_{mj} where L is needed, the near field of the source B_k obtained via the summation is forced to be the same as the true near field of B_k, obtained from the original integral operator. If there are q sample points in the quadrature rule, and q basis functions in the set, this leads to the square system of equations

$$\begin{bmatrix} B_1(t_{n1}) & B_1(t_{n2}) & \cdots & B_1(t_{nq}) \\ B_2(t_{n1}) & B_2(t_{n2}) & & B_2(t_{nq}) \\ \vdots & & & \vdots \\ B_q(t_{n1}) & B_q(t_{n2}) & \cdots & B_q(t_{nq}) \end{bmatrix} \begin{bmatrix} w_{n1} L(t_{mj}, t_{n1}) \\ w_{n2} L(t_{mj}, t_{n2}) \\ \vdots \\ w_{nq} L(t_{mj}, t_{nq}) \end{bmatrix} = \begin{bmatrix} \int_{\text{cell } n} B_1(t') \, K(t_{mj}, t') dt' \\ \int_{\text{cell } n} B_2(t') \, K(t_{mj}, t') dt' \\ \vdots \\ \int_{\text{cell } n} B_q(t') \, K(t_{mj}, t') dt' \end{bmatrix}. \tag{4.4}$$

This system can be solved for the numerical values of $L(t_{mj}, t_{n1})$ through $L(t_{mj}, t_{nq})$[1]. A similar system must be solved for each observer location mj in the near field of cell n. By creating L in this manner, the summation in (4.2) should produce accurate near fields for any current density $J_z(t)$ that can be adequately represented by the basis functions. Since the synthetic kernel L is only needed at a finite number of source/observer locations, and the system of (4.4) is only of order q, the required set of computations is relatively inexpensive. The most expensive part of the calculation is the accurate evaluation of the integrals appearing on the right-hand side of Equation (4.4) because of the kernel singularity.

In the electromagnetic equations of interest, the kernel of the integral operator is complex valued. In that situation, both the real and imaginary parts of the kernel are modified by the above procedure. Since it is only necessary to "correct" the non-analytic part of the original kernel, computational efficiency suggests that we separate the real and imaginary parts and apply the LCN procedure only to the non-analytic part. Although that is common practice, in the interest of simplicity, we present the development in the following sections as though we correct both parts.

---

[1] In practice, we solve for $\tilde{L}_{mjni} = w_{ni} L(t_{mj}, t_{ni})$, which are the actual matrix entries required in (4.12).

4.2   APPLICATION OF THE LCN TO THE MFIE

Consider the application of the LCN procedure to the MFIE for closed, perfectly conducting 2D scatterers (Chapter 3). For the TM polarization, the equation is given by

$$H_t^{\text{inc}}(t) = \frac{J_z(t)}{2} + \int_\Gamma J_z(t') \, C_1(t,t')dt' \tag{4.5}$$

where H_t^{inc} represents the known incident magnetic field, J_z is the unknown surface current density, t is a parametric variable along the contour Γ of the 2D structure,

$$C_1(t,t') = \frac{jk}{4} H_1^{(2)}(kR) \left\{ \frac{[y(t) - y(t')]}{R} \sin \Omega(t) + \frac{[x(t) - x(t')]}{R} \cos \Omega(t) \right\} \tag{4.6}$$

and

$$R = \sqrt{[x(t) - x(t')]^2 + [y(t) - y(t')]^2} \,. \tag{4.7}$$

The angle $\Omega(t)$ in (4.6) defines the orientation of the surface at location t with respect to the outward normal vector

$$\hat{n}(t) = \hat{x} \cos \Omega(t) + \hat{y} \sin \Omega (t) \,. \tag{4.8}$$

For a plane wave excitation, the incident field is

$$H_t^{\text{inc}}(t) = - \cos\{\theta^{\text{inc}} - \Omega(t)\}e^{-jk(x \cos \theta^{\text{inc}} + y \sin \theta^{\text{inc}})} \tag{4.9}$$

where θ^{inc} is the polar angle in the x-y plane into which the plane wave propagates.

The problem geometry is discretized into N cells, which in general may be curved and therefore defined parametrically using a mapping process such as that described in Appendix A. A q-point quadrature rule is required, in accordance with the conventional Nyström procedure, and for convenience, we employ Gauss-Legendre rules. Samples of the current density $J_z(t)$ are the primary unknowns to be determined, and the equation is enforced at each node of the quadrature rule, to produce a system of order Nq of the form

$$
\begin{bmatrix} H_{11}^{\text{inc}} \\ \vdots \\ H_{1q}^{\text{inc}} \\ H_{21}^{\text{inc}} \\ \vdots \\ H_{Nq}^{\text{inc}} \end{bmatrix}
=
\begin{bmatrix}
\left(\tfrac{1}{2} + Y_{11,11}\right) & \cdots & Y_{11,1q} & Y_{11,21} & \cdots & Y_{11,Nq} \\
\vdots & \ddots & \vdots & \vdots & & \vdots \\
Y_{1q,11} & \cdots & \left(\tfrac{1}{2} + Y_{1q,1q}\right) & Y_{1q,21} & \cdots & Y_{1q,Nq} \\
Y_{21,11} & \cdots & Y_{21,1q} & \left(\tfrac{1}{2} + Y_{21,21}\right) & \cdots & Y_{21,Nq} \\
\vdots & & \vdots & \vdots & \ddots & \vdots \\
Y_{Nq,11} & \cdots & Y_{Nq,1q} & Y_{Nq,21} & \cdots & \left(\tfrac{1}{2} + Y_{Nq,Nq}\right)
\end{bmatrix}
\begin{bmatrix} J_{11} \\ \vdots \\ J_{1q} \\ J_{21} \\ \vdots \\ J_{Nq} \end{bmatrix} . \tag{4.10}
$$

For source locations that are separated from the observer node by a distance greater than some chosen tolerance, the classical Nyström method is employed to construct the entries of the system of equations. Those entries are given by

$$\begin{aligned} Y_{mj,ni} &= w_i Q_{ni} C_1(t_{mj}, t_{ni}) \\ &= \frac{jk}{4} w_i Q_{ni} \left[\sin \Omega_m \frac{y_{mj} - y_{ni}}{R_{mj,ni}} + \cos \Omega_m \frac{x_{mj} - x_{ni}}{R_{mj,ni}} \right] H_1^{(2)}(k R_{mj,ni}) \end{aligned} \qquad (4.11)$$

where $R_{mj,ni} = \sqrt{[x_{mj} - x_{ni}]^2 + [y_{mj} - y_{ni}]^2}$, $x_{mj} = x(t_{mj})$, $y_{mj} = y(t_{mj})$, and where Q_{ni} is a sample of the Jacobian associated with the transformation used to define the curved cells at the location t_{ni} (Appendix A).

For cells that are closely-spaced, the matrix entries are given by the alternate values

$$Y_{mj,ni} = \tilde{L}_{mjni} = w_i L(t_{mj}, t_{ni}) \qquad (4.12)$$

which are obtained by solving Equation (4.4). The process requires a set of basis functions suitable for representing the current density. Reference [2] proposed the use of Legendre polynomials[2], and for illustration, we employ the first q polynomials as basis functions within (4.4). These basis functions, depicted in Figure 4.1, have the form

$$\left\{ 1, \ 2u - 1, \ 6u^2 - 6u + 1, \ 20u^3 - 30u^2 + 12u - 1, \ \ldots \right\} \qquad (4.13)$$

on a domain $0 \le u \le 1$ and form a hierarchical set. These functions are abruptly truncated at the cell boundaries, and there is no attempt to impose cell-to-cell continuity of the individual functions or the overall representation for the current density. The use of Legendre polynomials is consistent with the fact that those are the polynomials underlying the Gauss-Legendre quadrature rules; however, a wide variety of other basis functions could be used as alternatives.

The repeated solution of (4.4) for all the necessary observer locations $\{t_{mj}\}$ produces numerical values for $\tilde{L}$ that are used in (4.12). For example, if the locally-corrected kernel is used for the self cell ($m = n$) and the immediately adjacent cells, (4.4) must be solved $3Nq$ times in order to generate the necessary values in (4.12) to fill the system matrix. In practice, the "necessary" observer locations may be determined by their distance from the source; for instance, all entries where $R_{mj,ni}$ is less than a predetermined distance such as $\lambda/4$, where λ denotes the wavelength, may be replaced by the "corrected" values in (4.12).

Most of the required computation within these calculations is that used to evaluate the integrals on the right-hand side of (4.4); for the MFIE, these have the form

$$I_{mj,nk} = \frac{jk}{4} \int_{\text{cell } n} B_k(t') H_1^{(2)}(k R_{mj}) \left\{ \frac{[y_{mj} - y(t')]}{R_{mj}} \sin \Omega_{mj} + \frac{[x_{mj} - x(t')]}{R_{mj}} \cos \Omega_{mj} \right\} Q(t') dt' \qquad (4.14)$$

where $Q(t)$ is a Jacobian arising from the curved cell mapping, and

$$R_{mj} = \sqrt{[x_{mj} - x(t')]^2 + [y_{mj} - y(t')]^2} . \qquad (4.15)$$

---

[2]The monomials $\{1, u, u^2, \ldots\}$ could also be used; however, the Legendre polynomials are mutually orthogonal and produce a well-conditioned matrix in (4.4).

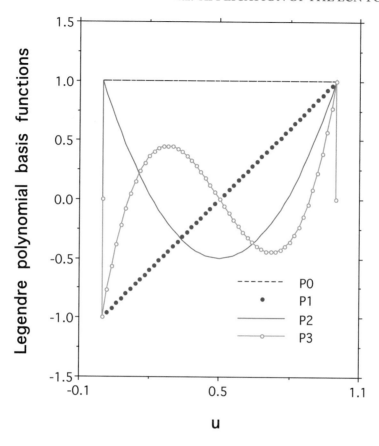

Figure 4.1: The first four Legendre polynomials defined on an interval $0 =\leq u \leq 1$.

Since the MFIE kernel is bounded, (4.14) can be evaluated by adaptive quadrature, with care taken for small R_{mj} to consider the limiting case of the kernel and avoid evaluating the Hankel function at $R_{mj} = 0$.

A comment on expected accuracy is in order. With the classical Nyström method and a smooth operator, the overall solution accuracy is determined by the quadrature rule (assuming that there are no other approximations that dominate the error). Although the MFIE integral operator can *not* be accurately integrated with Gauss-Legendre rules, for a smoother operator, the accuracy of the Nyström method should improve exponentially as the quadrature order is increased. And, since a q-point Gauss-Legendre rule can accurately integrate polynomials of degree $2q-1$, the resulting accuracy should be $O(h^{2q})$, where h is the characteristic cell dimension. With the LCN procedure, the synthetic kernel L limits the accuracy of the overall procedure to the degrees of freedom within the basis functions, which (in our implementation) are the first q Legendre polynomials (up to degree $q-$

1). Thus, while our expectation is that the LCN implementation will produce higher-order accuracy as q is increased, the error should behave as $O(h^q)$ and not as $O(h^{2q})$ as the cell dimension is reduced. Fundamentally, this limitation is due to the non-analytic nature of the integral operators associated with the EFIE and MFIE (which motivates the use of the LCN in the first place).

As an illustration, consider a circular cylinder of radius a, where $ka = 1$. Table 4.1 shows numerical values obtained for the closely-spaced matrix entries, with and without the local corrections. A 10-cell model was used. The numbers reported in Table 4.1 are the matrix entries for a source located at node 1 of cell 1, and a given observer location, for quadrature order $q = 3$. For the TM MFIE, which has a bounded kernel, there is little difference in the numbers obtained from the classical Nyström and LCN approaches (they agree to almost 4 decimal places). The LCN process embodied in Equation (4.4) creates a synthetic kernel that is almost identical to the original kernel, even in the source cell. For the MFIE, the answers exhibit little difference until the quadrature order q exceeds 3 and the error levels become quite small.

Table 4.1: Comparison of matrix entries for the TM MFIE for nodes on a 10-cell model of a circular cylinder with $ka = 1$, for quadrature order $q = 3$.

observer index (cell, node)	Nyström (3.25)–(3.27)	LCN (4.12)
1-1	0.48611 + j0.00000	0.48609 + j0.00000
1-2	−0.01471 + j0.00064	−0.01469 + j0.00064
1-3	−0.01596 + j0.00246	−0.01597 + j0.00246
2-1	−0.01657 + j0.00397	−0.01657 + j0.00397
2-2	−0.01712 + j0.00710	−0.01713 + j0.00710
2-3	−0.01690 + j0.01058	−0.01690 + j0.01058

Table 4.2 shows the error in the current density at the specular point on a circular cylinder with $ka = 3.5$, the same problem considered in Chapter 3. The TM MFIE is solved using the LCN with quadrature order $q = 3$, and Gauss-Legendre basis functions to correct the kernel samples whenever the observer and source cell centers were within $0.15\,\lambda$ of each other. Cells used within the cylinder model conform to the circular contour. The eigenfunction series was used as the exact reference solution [6]. As compared with the results in Table 3.1, those obtained with the LCN are far more accurate: the 30-cell result agrees with the exact solution to about 8 decimal places.

Figure 4.2 shows the error in the current density for a circular cylinder of radius $1\,\lambda$, where λ denotes the wavelength, as a function of the quadrature order q, for the TM MFIE. Gauss-Legendre basis functions were used with the LCN to correct the kernel samples whenever the observer was within $0.2\,\lambda$ of the source cell. Cells used within the cylinder model conformed to the exact circular boundary. The eigenfunction series was used as the reference solution [6], and Equation (3.22) was used for the error norm. As the number of unknowns increases, the error curves become essentially

Table 4.2: LCN MFIE results for the TM current density at the specular point ($\phi = 180$) on a circular cylinder with $ka = 3.5$ induced by a uniform plane wave with $\theta^{\text{inc}} = 0$. Results are obtained with a $q = 3$ Gauss-Legendre quadrature rule and cells that conform to the exact circular contour.

Number of cells	magnitude	phase (degrees)	Relative error
6	2.086	−166.3	1.1×10^{-2}
10	2.06814	−166.551	9.6×10^{-4}
20	2.0661704	−166.54889	1.1×10^{-6}
30	2.066168499	−166.5489217	3.4×10^{-8}
exact [6]	2.066168432	−166.5489226	

straight lines on the log-log scale, with an increasing slope as q increases. This confirms that the LCN process is successful at producing results that exhibit higher-order accuracy.

By comparing the error levels in Table 4.2 and Figure 4.2 to those presented in Chapter 3, it is clear that the LCN approach outperforms the classical Nyström method when applied to the MFIE. Figure 4.2 also illustrates the primary benefit of high order methods: with a higher value of q, the error curve exhibits a steeper slope. Therefore, one may obtain a given error level with fewer unknowns, or obtain a lower error level with the same number of unknowns. In fact, the error levels in both Table 4.2 and Figure 4.2 are decreasing at faster rates than the expected $O(h^q)$ (a phenomenon known as *superconvergence*).

While the current density in the Nyström and LCN approaches is defined by the samples at the nodes of the quadrature rule, it is often desired to compute values of the current at other locations. If this is necessary, an interpolation procedure based on the Legendre polynomials can be used to obtain the current density at any location. Specifically, for a representation involving q samples, the current in a given cell can be expressed in terms of $q = (p + 1)$ Legendre polynomials

$$J_z(t) \cong \alpha_0 P_0(t) + \alpha_1 P_1(t) + \cdots + \alpha_p P_p(t) \,. \tag{4.16}$$

By equating this expansion with the current samples at the nodes of the rule, we construct the q by q system

$$
\begin{bmatrix} J_z(t_{n1}) \\ J_z(t_{n2}) \\ \vdots \\ J_z(t_{nq}) \end{bmatrix}
=
\begin{bmatrix}
P_0(t_{n1}) & P_1(t_{n1}) & \cdots & P_p(t_{n1}) \\
P_0(t_{n2}) & P_1(t_{n2}) & & P_p(t_{n2}) \\
\vdots & & \ddots & \\
P_0(t_{nq}) & P_1(t_{nq}) & & P_p(t_{nq})
\end{bmatrix}
\begin{bmatrix} \alpha_0 \\ \alpha_1 \\ \vdots \\ \alpha_p \end{bmatrix} \,. \tag{4.17}
$$

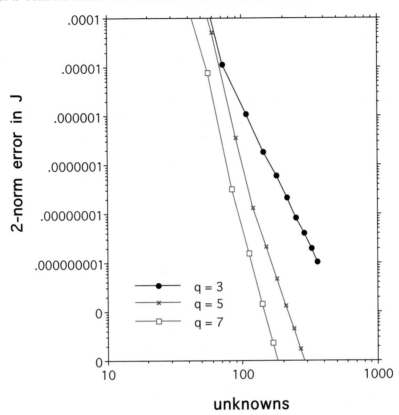

Figure 4.2: The error in the TM surface current density produced by the LCN method for a circular cylinder with $ka = 2\pi$ and quadrature rules of order $q = 3$, $q = 5$, and $q = 7$. Gauss-Legendre basis functions were used to correct all MFIE kernel samples whenever the observer was within $0.2\,\lambda$ of the source cell, where λ denotes the wavelength.

Inverting (4.17) yields the coefficients

$$
\begin{bmatrix} \alpha_0 \\ \alpha_1 \\ \vdots \\ \alpha_p \end{bmatrix} = \begin{bmatrix} P_0(t_{n1}) & P_1(t_{n1}) & \cdots & P_p(t_{n1}) \\ P_0(t_{n2}) & P_1(t_{n2}) & & P_p(t_{n2}) \\ \vdots & & \ddots & \\ P_0(t_{nq}) & P_1(t_{nq}) & & P_p(t_{nq}) \end{bmatrix}^{-1} \begin{bmatrix} J_z(t_{n1}) \\ J_z(t_{n2}) \\ \vdots \\ J_z(t_{nq}) \end{bmatrix}. \tag{4.18}
$$

Once the coefficients $\{\alpha_n\}$ are determined for a given cell, (4.16) may be used to obtain the current at any location within the cell or at the cell boundaries.

4.3   ALTERNATE INTERPRETATION OF THE LCN

The approximations inherent in the construction of the smoothed kernel L in the locally-corrected Nyström method are equivalent to the use of an explicit representation of the current density in terms of basis functions. Suppose that we consider an expansion of the current in cell n in terms of q basis functions

$$J_n(t) \cong \sum_{k=1}^{q} J_{nk} B_k(t) \tag{4.19}$$

where $\{J_{nk}\}$ are coefficients. The integral operator over cell n may, therefore, be approximated by the summation

$$\int_{\text{cell } n} J_n(t')K(t_{mj}, t')dt' \cong \sum_{k=1}^{q} J_{nk} \int_{\text{cell } n} B_k(t')K(t_{mj}, t')dt'. \tag{4.20}$$

Equation (4.20) was proposed by Tong and Chew [7] as an alternative way to implement the near-field interactions within the Nyström method although the idea was previously anticipated and studied by Gedney [8]. Although it may not be immediately obvious that this representation of the operator has any relationship to the LCN, we can demonstrate the equivalence between the two approaches by means of a "change of basis" procedure.

From the expansion in (4.19), samples of the current density at the quadrature nodes $\{t_{ni}\}$ are given by

$$\begin{bmatrix} J_n(t_{n1}) \\ J_n(t_{n2}) \\ \vdots \\ J_n(t_{nq}) \end{bmatrix} = \begin{bmatrix} B_1(t_{n1}) & B_2(t_{n1}) & \cdots & B_q(t_{n1}) \\ B_1(t_{n2}) & B_2(t_{n2}) & & B_q(t_{n2}) \\ \vdots & & \ddots & \\ B_1(t_{nq}) & B_2(t_{nq}) & & B_q(t_{nq}) \end{bmatrix} \begin{bmatrix} J_{n1} \\ J_{n2} \\ \vdots \\ J_{nq} \end{bmatrix}. \tag{4.21}$$

If we denote the q by q matrix in (4.21) by $\mathbf{M}$, we can express the conversion between coefficients of basis functions to the samples at quadrature nodes as

$$\begin{bmatrix} J_n(t_{n1}) \\ J_n(t_{n2}) \\ \vdots \\ J_n(t_{nq}) \end{bmatrix} = \mathbf{M} \begin{bmatrix} J_{n1} \\ J_{n2} \\ \vdots \\ J_{nq} \end{bmatrix} \tag{4.22}$$

while the inverse relation is given by

$$\begin{bmatrix} J_{n1} \\ J_{n2} \\ \vdots \\ J_{nq} \end{bmatrix} = \mathbf{M}^{-1} \begin{bmatrix} J_n(t_{n1}) \\ J_n(t_{n2}) \\ \vdots \\ J_n(t_{nq}) \end{bmatrix}. \tag{4.23}$$

It is suggestive that the matrix $\mathbf{M}$ is the transpose of the matrix that was used in (4.4) to construct the local corrections. Equation (4.4) can be rewritten as

$$
\mathbf{M}^{\mathrm{T}}
\begin{bmatrix}
w_{n1}L(t_{mj}, t_{n1}) \\
w_{n2}L(t_{mj}, t_{n2}) \\
\vdots \\
w_{nq}L(t_{mj}, t_{nq})
\end{bmatrix}
=
\begin{bmatrix}
\int_{\text{cell } n} B_1(t')K(t_{mj}, t')dt' \\
\int_{\text{cell } n} B_2(t')K(t_{mj}, t')dt' \\
\vdots \\
\int_{\text{cell } n} B_q(t')K(t_{mj}, t')dt'
\end{bmatrix} .
\tag{4.24}
$$

The connection between these ideas can be explored by considering the approximation of the integral operator over cell n by the summation in (4.20). From Equation (4.23), the coefficient J_{nk} can be expressed as

$$
J_{nk} = \sum_{i=1}^{q} (\mathbf{M}^{-1})_{ki} J(t_{ni}) .
\tag{4.25}
$$

Therefore, the summation on the right side of (4.20) can be written, without further approximation, as

$$
\begin{aligned}
I_{mjn} &= \sum_{k=1}^{q} J_{nk} \int_{\text{cell } n} B_k(t')K(t_{mj}, t')dt' \\
&= \sum_{k=1}^{q} \sum_{i=1}^{q} (\mathbf{M}^{-1})_{ki} J(t_{ni}) \int_{\text{cell } n} B_k(t')K(t_{mj}, t')dt' \\
&= \sum_{i=1}^{q} J(t_{ni}) \sum_{k=1}^{q} (\mathbf{M}^{-1})_{ki} \int_{\text{cell } n} B_k(t')K(t_{mj}, t')dt' \\
&= \sum_{i=1}^{q} J(t_{ni}) w_{ni} L(t_{mj}, t_{ni})
\end{aligned}
\tag{4.26}
$$

where Equation (4.24) is inverted and used to move from the third expression to the last expression, which is identical to Equation (4.2). This shows the equivalence between the use of an explicit basis set as in (4.20) and the LCN procedure employing those same basis functions to construct the corrected kernel. As a consequence, of this equivalence, the accuracy of the representation in (4.20) is the same as that of the LCN approach. The computational cost is also the same[3]. The equivalence was first reported by Gedney [8] and was also observed, indirectly, by Wildman and Weile [9].

4.4   APPLICATION OF THE LCN TO THE TM EFIE

The scattering of electromagnetic waves from conducting structures may also be posed in terms of an electric field integral equation (EFIE). For the TM polarization, the EFIE for a conducting cylinder

---

[3]Although there is no obvious advantage to the alternative interpretation for the scalar cases considered in this monograph, in the more complicated 3D vector situation, the use of explicit basis functions has been useful for facilitating LCN implementations.

has the form

$$E_z^{\text{inc}}(t) = jk\eta \int_\Gamma J_z(t') \frac{1}{4j} H_0^{(2)}(kR)dt' \tag{4.27}$$

where J_z is the unknown current density, t and t' are parametric variables defined on the contour Γ of the surface, $H_0^{(2)}$ is the zero order Hankel function of the second kind,

$$R = \sqrt{[x(t) - x(t')]^2 + [y(t) - y(t')]^2} \tag{4.28}$$

and where, for a plane wave excitation, the incident electric field (defined in the absence of the target) is

$$E_z^{\text{inc}}(t) = \eta e^{-jk(x \cos \theta^{\text{inc}} + y \sin \theta^{\text{inc}})} . \tag{4.29}$$

The cylinder surface may be discretized into N cells, which are typically curved and described by a parametric mapping involving a Jacobian $Q(t)$ (see Appendix A). In accordance with the classical Nyström method, a q-point Gauss-Legendre quadrature rule is used within each cell, and the primary unknowns to be determined are the current density samples at nodes of the quadrature rule. Following the steps of the Nyström method, we obtain a system of order Nq with the form

$$\mathbf{ZJ} = \mathbf{E}^i \tag{4.30}$$

where the entries of $\mathbf{J}$ are the unknowns $J_z(t_{ni})$, and the entries of $\mathbf{E}^i$ are samples of the incident field at the appropriate node locations.

For source/observer locations that are separated by a sufficient distance, the matrix entries are the Nyström samples

$$Z_{mj,ni} = w_i Q_{ni} \frac{k\eta}{4} H_0^{(2)}(kR_{mj,ni}) \tag{4.31}$$

where m denotes the observer cell, n the source cell, j the observer node, i the source node, and

$$R_{mj,ni} = \sqrt{[x_{mj} - x_{ni}]^2 + [y_{mj} - y_{ni}]^2} \tag{4.32}$$

where $x_{mj} = x(t_{mj})$, $y_{mj} = y(t_{mj})$, etc.

For closely-spaced separations, where the classical Nyström method is not applicable due to the singularity in the kernel, the LCN procedure is used to synthesize a "corrected" kernel. The entries in Equation (4.31) are replaced by the values

$$Z_{mj,ni} = w_i L(t_{mj}, t_{ni}) \tag{4.33}$$

where necessary values of L are obtained by the repeated solution of the system

$$\sum_{i=1}^q w_{ni} B_k(t_{ni}) L(t_{mj}, t_{ni}) = \frac{k\eta}{4} \int_{\text{cell } n} B_k(t') H_0^{(2)}(kR_{mj})Q(t')dt', \quad k = 1, 2, \ldots q \tag{4.34}$$

for each required value of m, where

$$R_{mj} = \sqrt{[x_{mj} - x(t')]^2 + [y_{mj} - y(t')]^2} \qquad (4.35)$$

and where, as in Section 4.2, the basis functions $\{B_k\}$ in (4.34) are subsectional Legendre polynomials. The integrals on the right-hand side of (4.34) must be evaluated with care due to the logarithmic singularity of the Hankel function. One way to accomplish this is through the use of the so-called *lin-log* quadrature rules, which will be discussed in Chapter 5. Other techniques for evaluating the TM EFIE integrals are discussed in Chapter 2 of [5].

To illustrate the transition from the Nyström samples in (4.31) for sufficiently separated points and the "locally corrected" kernel samples in (4.33) for closely-spaced points, Table 4.3 shows both for a 10-cell model of a circular cylinder of size $ka = 1$. The numbers reported in Table 4.3 are the matrix entries for a source located at node 1 of cell 1, and a given observer location, for quadrature order $q = 3$. The numbers are similar once the observer location is outside the source cell, but differ as the observer approaches the source. Beyond a separation distance of about 0.2 wavelengths, the numbers are essentially the same. (The real parts of the LCN matrix entries are based on the non-singular part of the kernel but are computed by the same local correction process used for the imaginary parts and reported in Table 4.3 for illustration. Those are the same as the Nyström samples right up to the singularity, confirming that the correction process works properly for non-singular kernels.)

Table 4.3: Comparison of matrix entries for the TM EFIE for nodes on a 10-cell model of a circular cylinder with $ka = 1$, for quadrature order $q = 3$.

observer index (cell, node)	Nyström (4.31)	LCN (4.33)
1-1	(infinite)	16.438 + j37.301
1-2	16.197 + j15.640	16.197 + j16.464
1-3	15.497 + j7.752	15.498 + j7.437
2-1	14.905 + j4.702	14.906 + j4.804
2-2	13.636 + j0.732	13.636 + j0.742
2-3	12.149 − j2.174	12.149 − j2.171
3-1	11.231 − j3.496	11.231 − j3.495
3-2	9.643 − j5.257	9.643 − j5.257
3-3	8.130 − j6.490	8.130 − j6.490

To demonstrate the accuracy of the LCN approach, Figure 4.3 shows the error in the current density for a circular cylinder of radius 1λ, as a function of the quadrature order q, obtained from the LCN procedure applied to the TM EFIE. Gauss-Legendre basis functions were used to correct all EFIE kernel samples whenever the observer was within $0.2\,\lambda$ of the source cell. Cells that conform

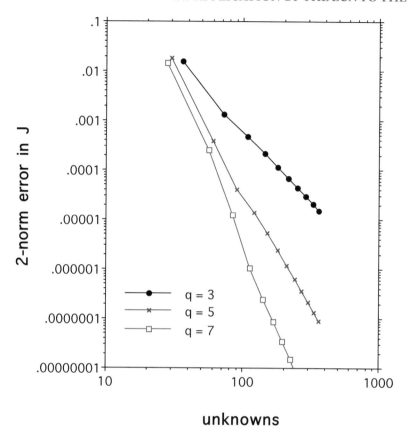

Figure 4.3: The error in the TM surface current density produced by the LCN method for a circular cylinder with $ka = 2\pi$ and quadrature rules of order $q = 3$, $q = 5$, and $q = 7$. Gauss-Legendre basis functions were used to synthesize alternate EFIE kernel samples whenever the observer was within $0.2\,\lambda$ of the source cell.

to the exact circular boundary were employed within the cylinder model. The error was determined by comparing the numerical results with the exact eigenfunction solution [6]. As the number of cells is increased with fixed q, the error decreases at a rate that approaches a straight line on the log-log scale. As q is increased, the slope of the error curves increases, confirming that the LCN results exhibit higher-order accuracy. For $q = 7$, the data in Figure 4.3 show that the accuracy of the current density improves by 7 orders of magnitude as the cell sizes are reduced by slightly more than one order of magnitude. This corresponds to an $O(h^q)$ behavior, where h denotes the cell size, which is the expected rate of error decrease.

In summary, an approach for discretizing the TM EFIE with the LCN procedure has been demonstrated. Since the EFIE kernel exhibits a logarithmic singularity, the classical Nyström method is not directly applicable to this equation. The LCN procedure is straightforward to implement and is able to produce high order behavior as q is increased.

4.5   APPLICATION OF THE LCN TO THE TE EFIE

For the TE polarization, the EFIE for scattering from conducting cylinders is

$$E_t^{inc}(t) = jk\eta \int_\Gamma \hat{t}(t) \bullet \hat{t}(t') \, J_t(t') \frac{1}{4j} H_0^{(2)}(kR) dt'$$
$$- \frac{\eta}{jk} \frac{\partial}{\partial t} \int_\Gamma \frac{\partial J_t}{\partial t'} \frac{1}{4j} H_0^{(2)}(kR) dt' \qquad (4.36)$$

where the primary unknown quantity is the transverse surface current density

$$\bar{J} = \hat{t}(t) J_t(t) \qquad (4.37)$$

defined in terms of the transverse unit vector $\hat{t}(t)$ previously introduced in Equation (3.6):

$$\hat{t}(t) = -\hat{x} \sin \Omega(t) + \hat{y} \cos \Omega(t) . \qquad (4.38)$$

The angle Ω is defined between the x-axis and the outward normal direction at points on the surface, and

$$R = \sqrt{[x(t) - x(t')]^2 + [y(t) - y(t')]^2} . \qquad (4.39)$$

The TE EFIE may be expressed in the equivalent form [10]

$$E_t^{inc}(t) = \int_\Gamma J_t(t) D(t, t') dt' \qquad (4.40)$$

where

$$D(t, t') = \frac{j\eta}{k} \left\{ k^2 \cos\left[\Omega(t) - \Omega(t')\right] + \sin \Omega(t) \sin \Omega(t') \frac{\partial^2}{\partial x^2} \right.$$
$$\left. - \sin\left[\Omega(t) + \Omega(t')\right] \frac{\partial^2}{\partial x \partial y} + \cos \Omega(t) \cos \Omega(t') \frac{\partial^2}{\partial y^2} \right\} \frac{1}{4j} H_0^{(2)}(kR) \qquad (4.41)$$

The kernel in (4.41) behaves as $O(R^{-2})$ for small arguments, indicating that the TE EFIE involves a hypersingular integral operator. Consequently, the classical Nyström method is not applicable to the TE EFIE. However, it should be possible to use the LCN procedure to generate numerical solutions of this equation.

The cylinder surface may be discretized into N cells, which may be curved and described by a parametric mapping. For ease of understanding, we omit the Jacobian functions from the following equations (see Appendix A for the corresponding equations with the Jacobians). Following the premise of the classical Nyström method, a q-point Gauss-Legendre quadrature rule with weights $\{w_i\}$ and nodes $\{t_{ni}\}$ is used within cell n, and the primary unknowns to be determined are the current density samples at the nodes of the quadrature rule. The Nyström approach produces a system of order Nq

$$\mathbf{ZJ} = \mathbf{E}^i \tag{4.42}$$

where the entries of $\mathbf{J}$ are the unknowns $J_t(t_{ni})$, and the entries of $\mathbf{E}^i$ are samples of the incident electric field at the appropriate node locations,

$$E^i_{mj} = \hat{t}(t_{mj}) \bullet \bar{E}^{\mathrm{inc}}(t_{mj}) . \tag{4.43}$$

For cells that are separated by a sufficient distance, the matrix entries are the Nyström samples

$$Z_{mj,ni} = w_i\, D(t_{mj}, t_{ni}) \tag{4.44}$$

where m denotes the observer cell, n the source cell, j the observer node, and i the source node.

For cells that are closely spaced, samples of the kernel are not suitable for use due to the $O(R^{-2})$ behavior of D. Instead, we employ

$$Z_{mj,ni} = w_i\, L(t_{mj}, t_{ni}) \tag{4.45}$$

where L is synthesized by solving the system

$$\sum_{i=1}^{q} w_{ni}\, B_k(t_{ni})\, L(t_{mj}, t_{ni}) = \int_{\mathrm{cell}\ n} B_k(t')\, D(t_{mj}, t')dt', \quad k = 1,\ 2,\ \ldots,\ q \tag{4.46}$$

at each observer location. As in the previous examples, the basis functions $\{B_k(t')\}$ are typically the first q Legendre polynomials as defined in Figure 4.1 and Equation (4.13). These basis functions are abruptly truncated at the cell edges, and the right-hand side of (4.46) must incorporate the jump discontinuities in the basis functions at the source cell edges, which do contribute to the result in this case because of the nature of the hypersingular operator.

The numerical evaluation of the integral

$$I_{mj,nk} = \int_{\mathrm{cell}\ n} B_k(t')\, D(t_{mj}, t')dt' \tag{4.47}$$

is somewhat more complicated than those encountered in the previous examples, due to the stronger singularity arising within the kernel D. There are several approaches that have been used in practice. These start by converting the integral back into the form in (4.36)

$$I_{mj,nk} = \frac{k\eta}{4} \int_{t'=\alpha}^{\beta} \hat{t}(t_{mj}) \bullet \hat{t}(t')\, B_k(t')\, G(R_{mj})dt'$$

$$+ \frac{k\eta}{4} \frac{1}{k^2} \left\{ \frac{\partial}{\partial t} \int_{t'=\alpha}^{\beta} \frac{\partial B_k}{\partial t'} G(R)dt' \right\}_{t=t_{mj}} \tag{4.48}$$

where $G(R) = H_0^{(2)}(kR)$,

$$R_{mj} = \sqrt{[x(t_{mj}) - x(t')]^2 + [y(t_{mj}) - y(t')]^2} \tag{4.49}$$

and where we have also provided explicit limits of integration for convenience. If the basis function discontinuities at the cell edges are taken into account, (4.48) can be written more completely as

$$
\begin{aligned}
I_{mj,nk} = & \frac{k\eta}{4} \int_{t'=\alpha}^{\beta} \hat{t}(t_{mj}) \bullet \hat{t}(t') \, B_k(t') \, G(R_{mj}) dt' \\
& + \frac{k\eta}{4} \frac{1}{k^2} \left\{ B_k(\alpha) \left. \frac{\partial G}{\partial t} \right|_{t=t_{mj}t'=\alpha} - B_k(\beta) \left. \frac{\partial G}{\partial t} \right|_{t=t_{mj}t'=\beta} \right\} \\
& + \frac{k\eta}{4} \frac{1}{k^2} \left\{ \frac{\partial}{\partial t} \int_{t'=\alpha}^{\beta} \frac{\partial B_k}{\partial t'} G(R) dt' \right\}_{t=t_{mj}} .
\end{aligned}
\tag{4.50}
$$

The integrals in (4.50) are not difficult to compute when the observer is not within the source cell. However, when the observer is located in the source cell, the first integral exhibits a logarithmic singularity and should be treated with lin-log quadrature rules or some other appropriate technique. The integral in the last term requires additional manipulation.

There are several possible approaches for the evaluation of the final term in (4.50) when the observer is within the source cell. We start with the observation that, for source and observer points t and t' that do not coincide, it is generally the case that

$$\left. \frac{\partial G}{\partial t'} \right|_{t,t'} \neq - \left. \frac{\partial G}{\partial t} \right|_{t,t'} . \tag{4.51}$$

However, it is always true that

$$\frac{\partial G}{\partial t'} = -\frac{\partial G}{\partial t} \text{ at } t = t' . \tag{4.52}$$

Therefore, the last expression in (4.50) may be manipulated using

$$\int_{t'=\alpha}^{\beta} \frac{\partial B_k}{\partial t'} \frac{\partial G}{\partial t} dt' = \int_{t'=\alpha}^{\beta} \left\{ \frac{\partial B_k}{\partial t'} \frac{\partial G}{\partial t} + f(t) \frac{\partial G}{\partial t'} \right\} dt' - f(t) \int_{t'=\alpha}^{\beta} \frac{\partial G}{\partial t'} dt' \tag{4.53}$$

where to convert the first integral on the right into a form that can be more easily evaluated, $f(t)$ must cancel $\partial B_k/\partial t'$ at $t = t'$. The obvious choice is

$$f(t) = \frac{\partial B_k}{\partial t} \tag{4.54}$$

where B_k is treated as a function of t, not of t'. The second integrand on the right side of (4.53) is a total derivative, which can be evaluated to obtain

$$\int_{t'=\alpha}^{\beta} \frac{\partial G}{\partial t'} dt' = G(t, \beta) - G(t, \alpha) . \tag{4.55}$$

Therefore, the final result can be expressed as

$$
\begin{aligned}
I_{mj,nk} = \frac{k\eta}{4} & \int_{t'=\alpha}^{\beta} \hat{t}(t_{mj}) \bullet \hat{t}(t') \, B_k(t') \, G(R_{mj}) dt' \\
& + \frac{k\eta}{4} \frac{1}{k^2} \left\{ B_k(\alpha) \frac{\partial G}{\partial t} \bigg|_{t=t_{mj},t'=\alpha} - B_k(\beta) \frac{\partial G}{\partial t} \bigg|_{t=t_{mj},t'=\beta} \right\} \\
& + \frac{k\eta}{4} \frac{1}{k^2} \frac{\partial B_k}{\partial t} \bigg|_{t=t_{mj}} \left\{ G(t_{mj},\alpha) - G(t_{mj},\beta) \right\} \\
& + \frac{k\eta}{4} \frac{1}{k^2} \int_{t'=\alpha}^{\beta} \left\{ \frac{\partial B_k}{\partial t'} \frac{\partial G}{\partial t} + \frac{\partial B_k}{\partial t} \frac{\partial G}{\partial t'} \right\} \bigg|_{t=t_{mj}} dt' \, .
\end{aligned}
\tag{4.56}
$$

The result in (4.56) is essentially the same as that proposed in [2]. The first and last terms involve integrals that contain logarithmic singularities and should be evaluated accordingly. As the observation point approaches the endpoints of the cell, several terms diverge. The expression can not be used when t_{mj} is located at an endpoint.

An alternate approach for evaluating (4.50) for the case when the observer resides within the source cell is to average the second integral over a small interval of length 2Δ, centered at the observation node t_{mj}, to obtain the approximate result [10, 11]

$$
\begin{aligned}
I_{mj,nk} \cong \frac{k\eta}{4} & \int_{t'=\alpha}^{\beta} \hat{t}(t_{mj}) \bullet \hat{t}(t') \, B_k(t') \, G(R_{mj}) dt' \\
& + \frac{k\eta}{4} \frac{1}{k^2} \left\{ B_k(\alpha) \frac{\partial G}{\partial t} \bigg|_{t=t_{mj},t'=\alpha} - B_k(\beta) \frac{\partial G}{\partial t} \bigg|_{t=t_{mj},t'=\beta} \right\} \\
& + \frac{k\eta}{4} \frac{1}{k^2} \frac{1}{2\Delta} \int_{t=t_{mj}-\Delta}^{t_{mj}+\Delta} \frac{\partial}{\partial t} \int_{t'=\alpha}^{\beta} \frac{\partial B_k}{\partial t'} G(R) dt' dt \, .
\end{aligned}
\tag{4.57}
$$

The last term is a total derivative and can be evaluated to yield

$$
\begin{aligned}
I_{mj,nk} \cong \frac{k\eta}{4} & \int_{t'=\alpha}^{\beta} \hat{t}(t_{mj}) \bullet \hat{t}(t') \, B_k(t') \, G(R_{mj}) dt' \\
& + \frac{k\eta}{4} \frac{1}{k^2} \left\{ B_k(\alpha) \frac{\partial G}{\partial t} \bigg|_{t=t_{mj},t'=\alpha} - B_k(\beta) \frac{\partial G}{\partial t} \bigg|_{t=t_{mj},t'=\beta} \right\} \\
& + \frac{k\eta}{4} \frac{1}{k^2} \frac{1}{2\Delta} \left\{ \int_{t'=\alpha}^{\beta} \frac{\partial B}{\partial t'} G(R_{mj+}) dt' - \int_{t'=\alpha}^{\beta} \frac{\partial B}{\partial t'} G(R_{mj-}) dt' \right\}
\end{aligned}
\tag{4.58}
$$

where

$$
R_{mj+} = \sqrt{[x(t_{mj}+\Delta) - x(t')]^2 + [y(t_{mj}+\Delta) - y(t')]^2}
\tag{4.59}
$$

$$
R_{mj-} = \sqrt{[x(t_{mj}-\Delta) - x(t')]^2 + [y(t_{mj}-\Delta) - y(t')]^2} \, .
\tag{4.60}
$$

The expression in (4.58) is similar to a "pulse tested" method-of-moments (MoM) matrix entry, as described in [5], and hence offers the convenience of using established procedures and legacy MoM software to evaluate the integrals. To maintain the linear independence of the equations in the case of multiple quadrature points per cell, and to mimic a direct sampling of the field so that the result in (4.45) smoothly merges with that of (4.44), it is suggested to use an interval size much smaller than the cell, perhaps as small as $\lambda/100$. For moderate accuracy requirements, experimentation shows little variation in the results as this interval size is perturbed [10, 11].

The approach in (4.57)–(4.58) is equivalent to using a finite difference approximation to the outer derivative in the last term of (4.48). As such, the process involves the cancellation of two large numerical values and is best done in a higher precision.

A third approach to the evaluation of the second integral in (4.48) was described in the appendix of [12] where an interpolation procedure was developed to permit a more accurate finite difference evaluation of the outer derivative.

To illustrate the transition from the Nyström samples to the locally corrected samples for closely-spaced points, Table 4.4 shows both for a 10-cell model of a circular cylinder of size $ka = 1$. The numbers reported in Table 4.4 are the matrix entries for a source located at node 1 of cell 1 and a given observer location for quadrature order $q = 3$. Compared to the TM EFIE, the TE numbers show a larger difference between the classical Nyström and LCN values as the observer approaches the source. It appears that the local correction region should extend to at least $0.2\,\lambda$ from the source.

Table 4.4: Comparison of matrix entries for the TE EFIE for nodes on a 10-cell model of a circular cylinder with $ka = 1$, for quadrature order $q = 3$.

observer index (cell, node)	Nyström (4.44)	LCN (4.45)
1-1	(infinite)	$8.22 - j\,1377.11$
1-2	$7.92 + j\,187.88$	$7.92 + j\,329.66$
1-3	$7.08 + j\,51.34$	$7.08 + j\,70.80$
2-1	$6.41 + j\,32.24$	$6.41 + j\,63.76$
2-2	$5.08 + j\,17.97$	$5.08 + j\,18.58$
2-3	$3.72 + j\,11.98$	$3.72 + j\,12.05$
3-1	$3.00 + j\,10.15$	$3.00 + j\,10.18$
3-2	$1.97 + j\,8.52$	$1.97 + j\,8.53$
3-3	$1.27 + j\,8.01$	$1.27 + j\,8.02$

The accuracy of the TE EFIE LCN approach is demonstrated in Figure 4.4, which shows the error in the current density for various models of a circular cylinder of radius $1\,\lambda$, for several values of q. Gauss-Legendre basis functions were used to synthesize alternate kernel samples whenever the observer was within $0.2\,\lambda$ of the source cell. Cells that conform to the exact circular boundary were employed within the cylinder model, and the interpolation method of [12] was used to evaluate

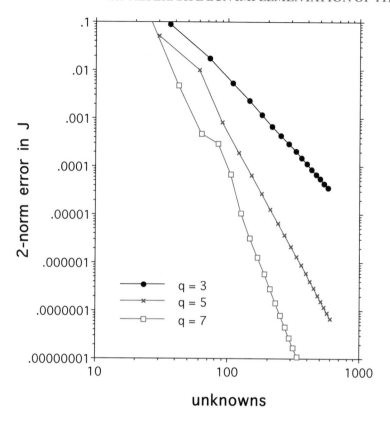

Figure 4.4: The error in the TE EFIE surface current density produced by the LCN method for a circular cylinder with $ka = 2\pi$ and quadrature rules of order $q = 3$, $q = 5$, and $q = 7$. Gauss-Legendre basis functions were used to synthesize alternate EFIE kernel samples whenever the observer was within $0.2\,\lambda$ of the source cell.

the last integral in (4.57). As the number of cells is increased with fixed q, the error decreases at rates that approach straight lines on the log-log scale. As q is increased, the slope of the error curves increases, confirming that the LCN results exhibit higher-order accuracy. For $q = 3, 5$, and 7, the data in Figure 4.4 appear to be following an $O(h^q)$ behavior, where h denotes the cell size, which is the expected rate of error decrease.

4.6   ALTERNATE LCN IMPLEMENTATION OF THE TE EFIE USING GAUSS-LOBATTO QUADRATURE

The preceding LCN implementations all employed Gauss-Legendre quadrature rules, and Legendre polynomials as basis functions to compute the locally-corrected kernel. In this section, we consider the use of Gauss-Lobatto rules to define the Nyström discretization. Gauss-Lobatto rules are closed, with nodes located at the endpoints of each interval, and thus permit the imposition of cell-to-cell current continuity. To synthesize the smoother kernel L for closely-spaced source/observer pairs, we employ Lagrangian polynomial basis functions [5]. These basis functions are interpolatory instead of hierarchical, but like Legendre polynomials, a set of q functions defined over a cell provides a representation that is complete to degree $p = q + 1$. Although when used as a basis, the Lagrangian functions provide first-order continuity between cells; we compute the locally-corrected kernel on a cell-by-cell basis and, therefore, sometimes have to incorporate a jump discontinuity at the cell edges.

The TE EFIE implementation is almost identical to that delineated in the preceding section, except that the Nyström samples in Equation (4.44) are obtained using the Lobatto nodes and weights; Lobatto nodes and weights are used in (4.46), and the basis functions in (4.46) are the Lagrangian polynomials. As an example, the three basis functions used for $q = 3$ on an interval $0 \leq u \leq 1$ are

$$B_1 = (1 - 2u)(1 - u) \tag{4.61}$$
$$B_2 = 4u(1 - u) \tag{4.62}$$
$$B_3 = u(2u - 1) . \tag{4.63}$$

The approximation in (4.57)–(4.58)) was used to evaluate the integrals on the right side of (4.46), with an interval size $2\Delta = \lambda/100$, since the technique used in (4.56) is not suitable when observer nodes are located at cell endpoints.

The primary difference in the implementation of the Lobatto LCN is that current density samples at the cell endpoints are common to both adjoining cells, and, therefore, the associated matrix entries have contributions from both cells. For instance, suppose cells v and m are adjacent, and node q of cell v and node 1 of cell m represent the same location. Then, the entry of the system matrix at that node (for a source some distance away) has a contribution from both observer locations:

$$Z_{m1,ni} = w_q D(t_{vq}, t_{ni}) + w_1 D(t_{m1}, t_{ni}) . \tag{4.64}$$

A similar approach is used for entries within the local correction footprint.

Figure 4.5 shows a comparison of the error produced by the Gauss-Lobatto LCN to that produced by the Gauss-Legendre LCN, for a circular cylinder with $ka = 3.5$, and representations of order $q = 3$ and $q = 5$. For a given number of unknowns, the accuracy of the two approaches is essentially identical for the same q value. Although the Legendre quadrature rules are usually more accurate than Lobatto rules, the accuracy of LCN implementations is limited by the functions used for the local corrections. Here, the Legendre polynomials and Lagrangian polynomials are complete to the same degree, and the underlying accuracy is apparently the same for the two approaches. Although the Lobatto representation imposes a de-facto cell-to-cell current continuity, that property does not improve the accuracy of the result on a "per unknown" basis.

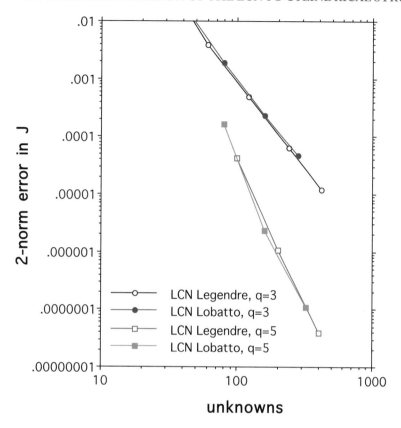

Figure 4.5: Error in the TE EFIE surface current density for a circular cylinder with ka = 3.5. The LCN method based on Legendre quadrature and Legendre polynomials (Section 4.5) is compared to one based on Lobatto quadrature and Lagrangian polynomials (Section 4.6).

4.7 INITIAL APPLICATION OF THE LCN TO CYLINDRICAL STRUCTURES WITH EDGE SINGULARITIES

The preceding examples have used circular cylinders for illustration, which is convenient since exact solutions are available for comparison. We now consider the scattering of a TM plane wave by a conducting cylinder whose cross section is triangular and investigate the solution of the MFIE using the LCN approach from Section 4.2. That formulation was based on Gauss-Legendre quadrature rules and employed Legendre polynomials as the basis functions to synthesize the corrected kernel. As mentioned in Chapter 3, a sharp corner in the cross section usually causes an edge singularity in the current density, and our primary purpose in considering the triangular geometry is to study how the Gauss-Legendre LCN approach performs when edges are present.

Specifically, consider a conducting cylinder with a cross section that is an equilateral triangle of side dimension 7λ, illuminated by a uniform TM plane wave incident symmetrically upon one corner of the cylinder. Cylinder models with 42, 84, 168, and 336 uniform cells were used, and locally-corrected samples of the MFIE kernel were used whenever the observer was within 0.6λ of the source cell. For illustration, we determine the error in the current density at 4 locations along the surface of the triangular cylinder, at positions 1.5λ, 2.0λ, 2.5λ, and 3.0λ from the illuminated edge. These locations form a subset of those for which a reference solution for the current density on this structure was published in [12]. These 4 locations all correspond to cell junctions, and the interpolation procedure in (4.16)–(4.18) was used within both cells adjacent to each location to obtain the current density.

Figure 4.6 shows the error obtained for $q = 2$, $q = 3$, and $q = 4$, averaged over the 8 extrapolated values at the 4 locations. As q increases, the error levels decrease but only slightly. Compared to the previous results for circular cylinders, these results exhibit much larger error levels and do not follow an $O(h^q)$ behavior — in fact, they do not even appear to decrease as fast as $O(h)$ for $q = 3$ and $q = 4$. The large error level is clearly a consequence of the presence of the edge singularities at the cylinder corners, which are not being properly modeled by the representation used in this LCN procedure. Subsequent chapters will extend the LCN procedure to handle corner singularities of this type.

4.8   SUMMARY

This chapter has presented the locally-corrected Nyström method, implemented with Gauss-Legendre quadrature rules and Legendre polynomial basis functions for the local corrections. The LCN approach was applied to the MFIE and EFIE, for two-dimensional problems of both the TM and TE polarizations. Results for circular cylinders exhibit high order behavior as the representation is improved and suggest that other smooth structures can be accurately analyzed using the LCN. In addition, an alternate implementation involving Gauss-Lobatto rules was investigated for solving the TE EFIE. The Lobatto approach, which imposes a form of cell-to-cell continuity on the representation for the surface currents, appears to produce results with the same accuracy (but not better) for a given number of unknowns as the Legendre-based LCN, which does not impose current continuity.

When applied to a triangular cylinder, for which the current density has edge singularities, the standard Legendre-based LCN discretization exhibits much larger error levels and does not appear to exhibit high order behavior. In order for the LCN approach to be useful for practical problems of interest, it must be extended to better handle edge singularities in the current density. Two aspects of the LCN procedure must be modified to accommodate the corner singularities: the quadrature rules on which the discretization is based and the basis functions used to implement the local corrections. Both of these require the identification of appropriate degrees of freedom to use in the representation of singular currents, and these are suggested by recent developments [12].

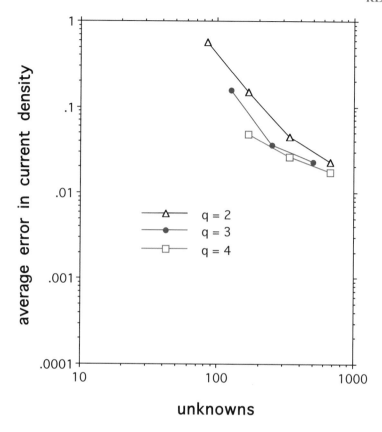

Figure 4.6: The error in the TM MFIE surface current density produced by the LCN method using Legendre quadrature and Legendre polynomials for a triangular cylinder of side dimension 7 λ. The error for $q = 2$, $q = 3$, and $q = 4$ is averaged over 4 locations near the illuminated edge of the cylinder.

Since quadrature rules capable of integrating singular functions are not widely tabulated, Chapter 5 will discuss their development. Chapter 6 will incorporate those quadrature rules and appropriate singular basis functions into the LCN approach.

REFERENCES

[1] S. D. Gedney, J. Ottusch, P. Petre, J. Visher, and S. Wandzura, "Efficient high-order discretization schemes for integral equation methods," *Digest of the 1997 IEEE Antennas and Propagation International Symposium*, Montreal, CA, pp. 1814–1817, July 1997.

[2] L. F. Canino, J. J. Ottusch, M. A. Stalzer, J. L. Visher, and S. M. Wandzura, "Numerical Solution of the Helmholtz Equation in 2D and 3D Using a High-Order Nyström Discretization," *J.*

Comp. Physics, vol. 146, pp. 627–663, 1998. DOI: 10.1006/jcph.1998.6077

[3] S. Kapur and D. E. Long, "High order Nyström schemes for efficient 3-D capacitance extraction," *Digest of Technical Papers of the IEEE/ACM International Conference on Computer-Aided Design (ICCAD 98)*, pp. 178–185, November 1998. DOI: 10.1145/288548.288604

[4] J. Strain, "Locally corrected multidimensional quadrature rules for singular functions," *SIAM J. Scientific Computing*, vol. 16, pp. 992–1017, 1995. DOI: 10.1137/0916058

[5] A. F. Peterson, S. L. Ray, and R. Mittra, *Computational Methods for Electromagnetics*. New York: IEEE Press, 1998.

[6] R. F. Harrington, *Time-harmonic Electromagnetic Fields*. New York: McGraw-Hill, 1961, pp. 232–235.

[7] M. S. Tong and W. C. Chew, "A higher-order Nyström scheme for electromagnetic scattering by arbitrarily shaped surfaces," *IEEE Antennas and Wireless Propagation Letters*, vol. 4, pp. 277–280, 2005. DOI: 10.1109/LAWP.2005.853000

[8] S. D. Gedney, "On deriving a locally-corrected Nyström scheme from a quadrature sampled moment method," *IEEE Trans. Antennas Propagat.*, vol. 51, pp. 2402–2412, September 2003. DOI: 10.1109/TAP.2003.816305

[9] R. A. Wildman and D. S. Weile, "Mixed-order testing functions on triangular patches for the locally corrected Nyström method," *IEEE Antennas and Wireless Propagation Letters*, vol. 5, pp. 370–372, 2006. DOI: 10.1109/LAWP.2006.881925

[10] A. F. Peterson, "Accuracy of currents produced by the locally-corrected Nyström method and the method of moments when used with higher-order representations," *Applied Computational Electromagnetics Society (ACES) Journal*, vol. 17, pp. 74–83, March 2002.

[11] A. F. Peterson, "Application of the locally-corrected Nyström method to the EFIE for the linear dipole," *IEEE Trans. Antennas Propagat.*, vol. 52, pp. 603–605, February 2004. DOI: 10.1109/TAP.2004.823955

[12] M. M. Bibby, A. F. Peterson, and C. M. Coldwell, "High order representations for singular currents at corners," *IEEE Trans. Antennas Propagat.*, vol. 56, pp. 2277–2287, August 2008. DOI: 10.1109/TAP.2008.926771

CHAPTER 5

Generalized Gaussian Quadrature

5.1 INTRODUCTION

Classical Gaussian quadrature rules are useful for analytic functions and for certain types of integrands containing singularities. In the present context, smooth or analytic functions are those whose derivatives are always bounded, while a singular function and/or its derivatives will at some point become unbounded. As an example, when there is only one singularity at an end of the domain, Jacobi rules [1] can be utilized. However, in practice it is sometimes desired to integrate a function with multiple singular terms at an end, or both ends, of the domain. Classical Gaussian quadrature rules are not efficient for such integrands although alternative quadrature rules for such functions are possible [2]. A general procedure for developing quadrature rules for integrands of the form

$$f(u) = g(u) + h(u)s(u) \tag{5.1}$$

where $g(u)$ and $h(u)$ are analytic functions and $s(u)$ is singular at one or both ends of the domain, was developed by Ma, Rokhlin, Wandzura, and Yarvin [3,4]. The content of this chapter is primarily based on their work.

The integrand $f(u)$ defined by (5.1) has the form of that arising in integral equations of electromagnetics, involving the product of a singular Green's function and an equivalent current density. For electromagnetic applications involving edges and corners, the integrands of interest contain more than one singular term since the current or charge density may also be singular at a corner. In these situations, we turn to what is now referred to as *generalized Gaussian quadrature*. Whereas the Jacobi-inspired rules use Legendre polynomials as basis functions, generalized Gaussian rules are generated using basis functions that have a 'natural' affinity for the integration needs at hand. The procedure for obtaining generalized Gaussian quadrature rules is illustrated below by obtaining the nodes and weights for the so-called "lin-log" rule [3], which has been employed for several formulations in Chapter 4.

5.2 EXAMPLE: DEVELOPMENT OF "LIN-LOG" RULES

Consider an integrand of the form of Equation (5.1), containing the set of terms

$$f(u) = \left(1, u, u^2, \ldots u^{n-1}\right) + \left(1, u, u^2, \ldots u^{n-1}\right) \ln u . \tag{5.2}$$

There are $2n$ terms in (5.2), and we seek a quadrature rule that can exactly integrate all $2n$ terms. (Such a rule will also exactly integrate any subset of those terms.) The rule will involve n nodes and weights and be of the form

$$\int_0^1 f(u)du = \sum_{i=1}^n w_i f(u_i) \,. \qquad (5.3)$$

The development of this rule will require the solution of $2n$ equations. Consider the case $n = 1$. For $n = 1$, the weights and nodes that will exactly integrate (5.2) are those that satisfy the two equations

$$\int_0^1 1 \, du = w_1 \qquad (5.4)$$

$$\int_0^1 \ln u \, du = w_1 \ln u_1 \,. \qquad (5.5)$$

The straightforward solution to these equations is $u_1 = e^{-1} \cong 0.368$, $w_1 = 1.0$. In other words, the integral of a function $f(u) = K_1 + K_2 \ln u$ over the unit interval is exactly given by the single term $(1) f(e^{-1}) = K_1 - K_2$.

For the case $n = 2$, the weights and nodes that will exactly integrate (5.2) must simultaneously satisfy four equations:

$$\int_0^1 1 \, du = w_1 + w_2 \qquad (5.6)$$

$$\int_0^1 \ln u \, du = w_1 \ln u_1 + w_2 \ln u_2 \qquad (5.7)$$

$$\int_0^1 u \, du = w_1 u_1 + w_2 u_2 \qquad (5.8)$$

$$\int_0^1 u \ln u \, du = w_1 u_1 \ln u_1 + w_2 u_2 \ln u_2 \,. \qquad (5.9)$$

The solution to this set of non-linear equations is decidedly less straightforward. One method to do this, proposed by [3], is to use a multi-dimensional Newton-Raphson approach. To facilitate this, the preceding equations are recast as:

$$f_1 = 1 - (w_1 + w_2) \qquad (5.10)$$
$$f_2 = -1 - (w_1 \ln u_1 + w_2 \ln u_2) \qquad (5.11)$$
$$f_3 = \frac{1}{2} - (w_1 u_1 + w_2 u_2) \qquad (5.12)$$
$$f_4 = -\frac{1}{4} - (w_1 u_1 \ln u_1 + w_2 u_2 \ln u_2) \qquad (5.13)$$

with the goal of obtaining the values of (u_1, u_2, w_1, w_2) that make $f_1 = f_2 = f_3 = f_4 = 0$. The definite integrals in (5.6)–(5.9) have been evaluated analytically, which is not only convenient, but necessary if high accuracy is to be achieved.

For a single variable, the Newton-Raphson root-finding method is based upon the Taylor series relation

$$f(u_{\text{new}}) = f(u_{\text{old}}) + f'(u_{\text{old}})(u_{\text{new}} - u_{\text{old}}) . \tag{5.14}$$

Since we seek the solution that produces $f(u_{\text{new}}) = 0$, the algorithm iterates on

$$u_{\text{new}} = u_{\text{old}} - \frac{f(u_{\text{old}})}{f'(u_{\text{old}})} \tag{5.15}$$

where u_{old} is a previous estimate, and u_{new} is a better approximation to the root of f. This process is repeated until $|u_{\text{new}} - u_{\text{old}}|$ is less than some predetermined limit.

For the system of four equations in (5.10)–(5.13), Equation (5.15) is recast in matrix form, leading to the system

$$\begin{bmatrix} \frac{\partial f_1}{\partial u_1} & \frac{\partial f_1}{\partial u_2} & \frac{\partial f_1}{\partial w_1} & \frac{\partial f_1}{\partial w_2} \\ \frac{\partial f_2}{\partial u_1} & \frac{\partial f_2}{\partial u_2} & \frac{\partial f_2}{\partial w_1} & \frac{\partial f_2}{\partial w_2} \\ \frac{\partial f_3}{\partial u_1} & \frac{\partial f_3}{\partial u_2} & \frac{\partial f_3}{\partial w_1} & \frac{\partial f_3}{\partial w_2} \\ \frac{\partial f_4}{\partial u_1} & \frac{\partial f_4}{\partial u_2} & \frac{\partial f_4}{\partial w_1} & \frac{\partial f_4}{\partial w_2} \end{bmatrix}_{(u_1,u_2,w_1,w_2)_{\text{old}}} \begin{bmatrix} \Delta u_1 \\ \Delta u_2 \\ \Delta w_1 \\ \Delta w_2 \end{bmatrix} = - \begin{bmatrix} f_1 \\ f_2 \\ f_3 \\ f_4 \end{bmatrix}_{(u_1,u_2,w_1,w_2)_{\text{old}}} \tag{5.16}$$

where the entries are specified by an estimate $(u_1, u_2, w_1, w_2)_{\text{old}}$ of the solution. At each step of the iterative process, these equations are solved (by LU factorization, singular value decomposition, or some other means), to determine a new column vector of corrections, which are used to update the solution estimate according to

$$\begin{bmatrix} u_1 \\ u_2 \\ w_1 \\ w_2 \end{bmatrix}_{\text{new}} = \begin{bmatrix} u_1 \\ u_2 \\ w_1 \\ w_2 \end{bmatrix}_{\text{old}} + \begin{bmatrix} \Delta u_1 \\ \Delta u_2 \\ \Delta w_1 \\ \Delta w_2 \end{bmatrix} . \tag{5.17}$$

For the specific equations above, we obtain the system

$$\begin{bmatrix} 0 & 0 & 1 & 1 \\ \frac{w_1}{u_1} & \frac{w_2}{u_2} & \ln u_1 & \ln u_2 \\ w_1 & w_2 & u_1 & u_2 \\ w_1(1+\ln u_1) & w_2(1+\ln u_2) & u_1 \ln u_1 & u_2 \ln u_2 \end{bmatrix}_{(u_1,u_2,w_1,w_2)_{\text{old}}} \begin{bmatrix} \Delta u_1 \\ \Delta u_2 \\ \Delta w_1 \\ \Delta w_2 \end{bmatrix} = \begin{bmatrix} f_1 \\ f_2 \\ f_3 \\ f_4 \end{bmatrix}_{(u_1,u_2,w_1,w_2)_{\text{old}}} . \tag{5.18}$$

To apply the iterative procedure, one requires an initial estimate $(u_1, u_2, w_1, w_2)_{old}$. This estimate is critical to the convergence of the algorithm, as the Newton-Raphson procedure will often diverge if the starting point is too far from the solution. In addition, some form of relaxation is usually used to initially slow the convergence. Specifically, (5.17) is replaced by

$$
\begin{bmatrix} u_1 \\ u_2 \\ w_1 \\ w_2 \end{bmatrix}_{new} = \begin{bmatrix} u_1 \\ u_2 \\ w_1 \\ w_2 \end{bmatrix}_{old} + \alpha \begin{bmatrix} \Delta u_1 \\ \Delta u_2 \\ \Delta w_1 \\ \Delta w_2 \end{bmatrix} \tag{5.19}
$$

where α is a scaling factor that satisfies $0 < \alpha \leq 1$. For instance, $\alpha = 1/8$ might be used for the first pass through the Newton-Raphson procedure, followed by $\alpha = 1/4$, $\alpha = 1/2$, and then finally $\alpha = 1$ on all subsequent passes. The iterative process continues until the maximum absolute value of each term in the correction vector is less than a desired error level.

Ma et al. suggest procedures for determining the initial estimate [3]. In this chapter, we offer a slightly different approach, which is to determine the starting point for an n-point rule by taking the centers of the intervals defined by the previous $(n-1)$ nodes as the new initial node estimates. This requires that we develop an entire family of rules from a 1-point rule on up. For the weights, the initial estimate of $1/n$ appears adequate at each level of this process. We previously determined that the solution of the 1-point rule in (5.4)–(5.5) is $u_1 = 0.368$, $w_1 = 1.0$. In our approach, we average the single node with the endpoint values (0 and 1) to obtain initial estimates for $n = 2$ of

$$
u_1 = \left(\frac{0 + 0.368}{2} \right) = 0.184 \tag{5.20}
$$

$$
u_2 = \left(\frac{0.368 + 1}{2} \right) = 0.684 . \tag{5.21}
$$

For the weights, we use $w_1 = w_2 = 1/2$. After substituting these values into (5.18), and solving, we obtain a correction vector that is used to update the estimate in (5.19). For this problem, 6 iterations are required to obtain nodes and weights for the 2-point rule that are accurate to 7 digits. After 8 iterations, the results are accurate to about 15 decimal places. For this example, the final values to 7 digits are $u_1 = 0.0882969$, $w_1 = 0.2984999$, $u_2 = 0.6751865$, and $w_2 = 0.7015001$.

For $n = 3$, we proceed to construct an order-6 system of the form of (5.18), and initiate the iteration with the estimates

$$u_1 = \left(\frac{0 + 0.0882969}{2}\right) \cong 0.044 \tag{5.22}$$

$$u_2 = \left(\frac{0.0882969 + 0.6751865}{2}\right) \cong 0.382 \tag{5.23}$$

$$u_3 = \left(\frac{0.6751865 + 1}{2}\right) \cong 0.838 \tag{5.24}$$

and $w_1 = w_2 = w_3 = 1/3$. As the number of equation pairs grows so does the number of passes through the iterative loop and the condition number of the Jacobian matrix in (5.18). For just ten nodes, in order to get answers that have an error level approximately equal to machine epsilon in single (10^{-7}), double (10^{-16}) and quad precision (10^{-34}) the number of passes through the iterative loop was 8, 9 and 10, respectively. The matrix condition number was approximately 10^{13} in all cases. Because the condition number grows rapidly as the value of n grows, it is necessary to use multi-precision software [5] or some comparable technique to ensure adequate precision in the computations.

5.3   HIGH ORDER REPRESENTATION OF CURRENT DENSITY AT EDGES IN TWO-DIMENSIONS

Recent developments in two-dimensional electromagnetics formulations suggest a methodology for extending the LCN method in order to treat edge singularities [6,7]. Currents near a geometrical edge follow the asymptotic behavior of the classical infinite wedge solution [8]. A general asymptotic expression for the current density as a function of ρ on the face of the wedge with interior angle α, near the tip ($\rho = 0$), can be written for the transverse-magnetic (TM)-to-z case as

$$J_z \sim \sum_{m=0}^{\infty} \sum_{n=1}^{\infty} c_{mn} \rho^{2m+\upsilon_n-1} \tag{5.25}$$

where a cylindrical coordinate system (ρ, ϕ, z) is employed. In (5.25),

$$\upsilon_n = \frac{n\pi}{(2\pi - \alpha)}, \quad n = 1,\ 2,\ 3,\ \dots\ . \tag{5.26}$$

A similar expression for the transverse electric (TE)-to-z case is

$$J_\rho \sim \sum_{m=0}^{\infty} \sum_{n=0}^{\infty} d_{mn} \rho^{2m+\upsilon_n} \tag{5.27}$$

where v_n is defined as

$$v_n = \frac{n\pi}{(2\pi - \alpha)}, \quad n = 0, 1, 2, \ldots . \tag{5.28}$$

References [6,7] proposed a hierarchical family of basis functions for representing the current density in cells adjacent to geometric corners. For cells that are not adjacent to a corner of the contour, a Legendre expansion of order q was employed. In the corner cells, the same representation is augmented by including some number of terms with non-integer exponents from (5.25) or (5.27). There is a trade-off between the number of singular terms used in corner cells, and the relative dimension of the corner cell. Reference [7] concluded that, for a wide range of problems, a nearly optimum solution was obtained with corner cells that were twice the dimension of the other cells in the model and a representation that incorporated a number of singular terms equal to the number of regular terms. That approach will be used in Chapter 6 to implement singular representations into the LCN approach.

As an example, for a conducting wedge with interior angle of 90 degrees, the transverse-magnetic-to-z (TM) current near the corner involves exponents, in increasing order, of

$$\left\{ -\frac{1}{3}, \frac{1}{3}, 1, \frac{5}{3}, \frac{7}{3}, 3, \frac{11}{3}, \ldots \right\}. \tag{5.29}$$

The methodology of [7] can be implemented for a target containing a 90 degree corner by constructing basis functions with the combination of exponents as summarized in Table 5.1.

Table 5.1: Exponents of the degrees of freedom used for a given order, for cells near a 90 degree corner (TM case). The corner cells involve twice as many degrees of freedom as the other cells, and are twice as large.

Order of representation q	Exponents used in non-corner cells	Exponents used in corner cells
1	0	−1/3, 0
2	0, 1	−1/3, 0, 1/3, 1
3	0, 1, 2	−1/3, 0, 1/3, 1, 5/3, 2
4	0, 1, 2, 3	−1/3, 0, 1/3, 1, 5/3, 2, 7/3, 3
5	0, 1, 2, 3, 4	−1/3, 0, 1/3, 1, 5/3, 2, 7/3, 3, 11/3, 4

Two important aspects of an LCN approach necessary to implement the singular representation described above are a set of basis functions that can be used to synthesize a "locally corrected" kernel when the sources reside in a cell containing a corner singularity and a family of quadrature rules that can integrate those singular basis functions to high accuracy. The basis functions are obtained from the degrees of freedom identified above (those in column 3 of Table 5.1) and a Gram-Schmidt orthogonalization process. These will be presented in Chapter 6. The quadrature rules need to be synthesized from the same set of degrees of freedom, using the procedure described in Section 5.2.

Since a q-point quadrature rule can exactly integrate $2q$ independent terms, additional degrees of freedom from the family of terms for the relevant corner angle were employed in the generation of the quadrature rules. For simplicity, we use the same terms appearing in the basis function of order $2q$. It is expected that the accuracy of the overall LCN analysis will be limited by the basis functions used within the local correction procedure, and not by the quadrature rules.

5.4 QUADRATURE RULES FOR THE SINGULAR DEGREES OF FREEDOM IN TABLE 5.1

The process of developing generalized Gaussian quadrature rules capable of integrating the degrees of freedom in Table 5.1 is similar to that outlined in Section 5.2 for the lin-log rule. For $q = 1$, the LCN requires a 2-point quadrature rule capable of integrating the two desired degrees of freedom $(u^{-1/3}$ and $1)$. Since the generation of a 2-point rule requires 4 degrees of freedom, we somewhat arbitrarily select an integrand of the form

$$f(u) = \alpha u^{-1/3} + \beta + \gamma u^{1/3} + \delta u \tag{5.30}$$

containing the additional degrees of freedom $u^{1/3}$ and u from column 3 of Table 5.1. (The coefficients α, β, etc., in (5.30) are arbitrary and do not enter into the process described below.) For $q = 2$, the proposed representation involves a 4-point rule. To generate that rule, 8 terms are required, and we select a set with the exponents

$$\left\{ -\frac{1}{3},\ 0,\ \frac{1}{3},\ 1,\ \frac{5}{3},\ 2,\ \frac{7}{3},\ 3 \right\}. \tag{5.31}$$

The set in Equation (5.31) has the desired degrees of freedom $(u^{-1/3},\ 1,\ u^{1/3},\ u)$ and the additional functions $(u^{5/3},\ u^2,\ u^{7/3},\ u^3)$.

To illustrate the process for these functions, we start with a 1-point rule for the two degrees of freedom $(1,\ u^{-1/3})$ and impose Equation (5.3) for each term to obtain

$$f_1 = 0 = \int_0^1 u^{-1/3}\, du - w_1 (u_1)^{-1/3} \tag{5.32}$$

$$f_2 = 0 = \int_0^1 1\, du - w_1. \tag{5.33}$$

This system has the solution $u_1 = 0.296296$, $w_1 = 1$. The next set of equations to solve is the set for the degrees of freedom in (5.30):

$$f_1 = 0 = \int_0^1 u^{-1/3}\, du - w_1(u_1)^{-1/3} - w_2(u_2)^{-1/3} \tag{5.34}$$

$$f_2 = 0 = \int_0^1 1\, du - w_1 - w_2 \tag{5.35}$$

$$f_3 = 0 = \int_0^1 u^{1/3}\, du - w_1(u_1)^{1/3} - w_2(u_2)^{1/3} \tag{5.36}$$

$$f_4 = 0 = \int_0^1 u\, du - w_1 u_1 - w_2 u_2 \, . \tag{5.37}$$

Here, the starting values for the Newton-Raphson procedure are calculated as indicated above, namely

$$u_1 = \left(\frac{0 + 0.296}{2}\right) = 0.148 \tag{5.38}$$

$$u_2 = \left(\frac{0.296 + 1}{2}\right) = 0.648 \tag{5.39}$$

with $w_1 = w_2 = 1/2$. Six iterations of Equation (5.16) are required to obtain results that are accurate to approximately 7 decimal places, with 13 digits obtained after 7 iterations. The results are $u_1 = 0.0508577$, $w_1 = 0.2154983$, $u_2 = 0.6233770$, and $w_2 = 0.7845017$.

We now add two more equations to represent the first 6 terms in (5.31), resulting in a total of six equations to solve. The new starting estimate of the solution is

$$u_1 = \left(\frac{0 + 0.0598}{2}\right) \cong 0.030 \tag{5.40}$$

$$u_2 = \left(\frac{0.0598 + 0.6420}{2}\right) \cong 0.351 \tag{5.41}$$

$$u_3 = \left(\frac{0.6420 + 1}{2}\right) \cong 0.821 \tag{5.42}$$

with $w_1 = w_2 = w_3 = 1/3$, and the six equations are:

$$f_1 = 0 = \int_0^1 u^{-1/3}\, du - w_1(u_1)^{-1/3} - w_2(u_2)^{-1/3} - w_3(u_3)^{-1/3} \tag{5.43}$$

$$f_2 = 0 = \int_0^1 1\, du - w_1 - w_2 - w_3 \tag{5.44}$$

$$f_3 = 0 = \int_0^1 u^{1/3}\, du - w_1(u_1)^{1/3} - w_2(u_2)^{1/3} - w_3(u_3)^{1/3} \tag{5.45}$$

$$f_4 = 0 = \int_0^1 u\, du - w_1 u_1 - w_2 u_2 - w_3 u_3 \tag{5.46}$$

$$f_5 = 0 = \int_0^1 u^{5/3}\, du - w_1(u_1)^{5/3} - w_2(u_2)^{5/3} - w_3(u_3)^{5/3} \tag{5.47}$$

$$f_6 = 0 = \int_0^1 u^2\, du - w_1(u_1)^2 - w_2(u_2)^2 - w_3(u_3)^2 \,. \tag{5.48}$$

These equations are recast in a form similar to (5.16), and solved by the iterative process, to produce the 3-point nodes and weights. Those parameters are subsequently used to start the iteration on the order-8 matrix for the 4-point rule, and so on. This procedure yields nodes and weights of any order for the family of terms in (5.29). Our treatment of edge singularities with the LCN method in Chapter 6 requires only the even-order rules. These nodes and weights are tabulated in Appendix B, along with those for several other wedge angles.

5.5   SUMMARY

This chapter has presented a procedure for developing generalized Gaussian quadrature rules for integrating singular functions arising in electromagnetics problems. These functions may contain multiple singular terms, such as those associated with the current density near target edges. Quadrature rules of this type will be used in Chapter 6 to extend the LCN implementations of Chapter 4 to scatterers with edges.

REFERENCES

[1] W. H. Press, S. A. Teukolsky, W. T. Vetterling, and B. P. Flannery, *Numerical Recipes in Fortran*, Cambridge University Press, 1992, pp. 144–149.

[2] S. Karlin and W. J. Studden, *Tchebycheff Systems: With Applications in Analysis and Statistics*, New York: Wiley, 1966.

[3] J. H. Ma, V. Rokhlin, and S. Wandzura, "Generalized Gaussian quadrature rules for systems of arbitrary functions," *SIAM J. Numer. Anal.*, vol. 33, pp. 971–996, June 1996. DOI: 10.1137/0733048

[4] N. Yarvin and V. Rokhlin, "Generalized Gaussian quadratures and singular value decompositions of integral operators," *SIAM J. Sci. Computing*, vol. 20, pp. 699–718, 1998. DOI: 10.1137/S1064827596310779

[5] D. H. Bailey, "A Fortran-90 Based Multi-precision System", *ACM Trans. on Mathematical Software*, Vol. 20, No. 4, pp. 379–387, Dec 1995. See also *RNR Technical Report RNR-90-022*, 1993 and http://crd.lbl.gov/~dhbailey/mpdist/.

[6] M. M. Bibby, A. F. Peterson, and C. M. Coldwell, "High order representations for singular currents at corners," *IEEE Trans. Antennas Propagat.*, vol. 56, pp. 2277–2287, August 2008. DOI: 10.1109/TAP.2008.926771

[7] M. M. Bibby, A. F. Peterson, and C. M. Coldwell, "Optimum cell size for high order singular basis functions at geometric corners," *ACES Journal*, vol. 24, pp. 368–374, August 2009.

[8] R. F. Harrington, *Time Harmonic Electromagnetic Fields*. New York: McGraw-Hill, 1961, pp. 238–242.

CHAPTER 6

LCN Treatment of Edge Singularities

Singularities in the current or charge density arise at sharp edges or corners of objects. Section 5.3 outlined an approach for representing the current density near such an edge, based on recent investigations that were developed and tested in the context of the method of moments discretization of integral equations [1,2]. In this chapter, we implement a similar capability into the LCN method. Most approaches for treating edge singularities only allow for a single singular term in the representation, and previous studies of that type have been carried out for the LCN method [3,4]. In contrast, our procedure permits the incorporation of multiple singular terms in order to obtain true high order behavior. Two features are necessary to accurately model edge singularities with the LCN: (1) basis functions incorporating the specific edge singularities, for use in synthesizing a "corrected" kernel, and (2) quadrature rules capable of integrating the singular degrees of freedom. The approach will be illustrated by several examples.

6.1 TM SCATTERING FROM A TRIANGULAR CYLINDER, REVISITED

The naïve treatment of scattering from a cylinder, whose cross section is an equilateral triangle, was briefly considered in Chapter 4, based on the LCN with Gauss-Legendre quadrature rules and Legendre polynomial basis functions for local corrections. Since the underlying representation of the current density did not make any attempt to incorporate the singularities at the cylinder edges, the results (Figure 4.6) did not improve substantially as q was increased and did not exhibit high order behavior.

We now return to that problem, but consider a more appropriate representation of the current in cells adjacent to an edge. For the TM case, the current near a 60 degree corner has the asymptotic form given in Equations (5.25)–(5.26), and it includes exponents from the set

$$\left\{ \frac{-2}{5}, \frac{1}{5}, \frac{4}{5}, \frac{7}{5}, \frac{8}{5}, 2, \frac{11}{5}, \frac{13}{5}, \cdots \right\}. \tag{6.1}$$

Following the approach of [1,2], we employ a representation where regular degrees of freedom (integer exponents) and singular degrees of freedom (fractional exponents) are combined in a 1:1 ratio within cells adjacent to an edge. Table 6.1 illustrates the specific degrees of freedom as a function of order q. Reference [2] concluded that to optimize this representation, these "corner" cells should be twice as large as the other cells in the model. Consequently, there will be twice as many unknowns in each corner cell, but those cells will be twice as large as non-corner cells, so the overall density of unknowns is the same.

Table 6.1: Example: Exponents of the degrees of freedom used for a given order, for cells near a 60 degree corner (TM case). The corner cells involve twice as many degrees of freedom as the other cells, and are twice as large.

Order of representation q	Exponents used in non-corner cells	Exponents used in corner cells
1	0	$-2/5, 0$
2	0, 1	$-2/5, 0, 1/5, 1$
3	0, 1, 2	$-2/5, 0, 1/5, 1, 4/5, 2$
4	0, 1, 2, 3	$-2/5, 0, 1/5, 1, 4/5, 2, 7/5, 3$
5	0, 1, 2, 3, 4	$-2/5, 0, 1/5, 1, 4/5, 2, 7/5, 3, 8/5, 4$

The exponents in the third column of Table 6.1 can be used to construct a set of hierarchical, orthogonal basis functions, using a Gram-Schmidt process to impose the orthogonality

$$\int_0^1 B_i(u) B_j(u)\, du = 0, \quad i \neq j. \tag{6.2}$$

The first six members of this set are given in Section B.5 of Appendix B. These functions are to be used as the basis set when constructing the "locally corrected" kernel L for near-field interactions, for sources in the cells adjacent to a 60 degree corner. For $q=1$, functions B_1 and B_2 would be used; for $q = 2$, B_1–B_4 would be used, and all six would be used for $q = 3$.

The basic LCN discretization procedure is as follows. The triangular cylinder contour is divided into cells, with the corner cells twice as large as the others. In non-corner cells, a Gauss-Legendre quadrature rule of order q is employed, and the current density samples at the nodes of the rule are the unknowns (q per cell). For observer nodes located a sufficient distance away from the source cell, a traditional Nyström sampling of the kernel K is used, with weights obtained from the appropriate Gauss-Legendre rule. For observer nodes closely spaced to source nodes in non-corner cells, Legendre polynomial basis functions in the source cell are used to synthesize a new kernel L (exactly as previously described in Chapter 4 — Equation (4.4) is solved to determine the samples of L).

In the corner cells, we use the nodes of the $2q$-point generalized Gaussian quadrature rule from Section B.5 of Appendix B to define the current density samples ($2q$ unknowns per corner cell). For observer nodes located a sufficient distance away from the source cell, a traditional Nyström sampling of the kernel K is used, with weights obtained from the generalized Gaussian quadrature rule (appropriately oriented so that the singularity at $u = 0$ is positioned at the cylinder corner). For observer nodes within or closely spaced to the corner cells, the singular basis functions from

Section B.5 are used to synthesize a new kernel L according to Equation (4.4). The right-hand side of (4.4) involves an integration over the orthogonal singular basis functions; appropriate generalized Gaussian quadrature rules can be developed to assist in the accurate evaluation of that integral.

To illustrate the approach, consider the scattering of a TM plane wave from a cylinder whose cross section is an equilateral triangle of $21\,\lambda$ perimeter where λ denotes the wavelength (the example from Section 4.7). The uniform plane wave is incident symmetrically upon one corner. The numerical solution is obtained by an application of the LCN approach to the magnetic field integral equation (MFIE). As mentioned previously, to implement the optimum approach from [2], the cells adjacent to corners are twice as large as other cells and use twice as many degrees of freedom (a number of singular terms equal to the number of regular terms). A high accuracy reference solution for the induced currents on the $21\,\lambda$ cylinder was presented in Table I of [1] and is used for comparison.

Figure 6.1 shows the current density error versus the number of unknowns, for representations of order q equal to 2, 3, and 4, and cylinder models with 12, 26, and 54 cells per side of the triangle. Because the corner cells are relatively large ($1\,\lambda$ for the coarsest model), locally-corrected samples of the MFIE kernel were used whenever the source and observer cell centers were within $0.6\,\lambda$ of each other. The plots show the error in the current density averaged over four locations along the surface, at distances of $1.5\,\lambda$, $2.0\,\lambda$, $2.5\,\lambda$, and $3.0\,\lambda$ from the illuminated corner of the triangular cylinder. The error is calculated according to (3.22). These locations form a subset of those for which a reference solution for the current density on this structure was published in [1]. These 4 locations all correspond to cell junctions, and the interpolation procedure in (4.16)–(4.18) was used within both cells adjacent to each location to obtain the current density. These error curves, in solid lines, closely follow trends of $O(h^2)$, $O(h^3)$, and $O(h^4)$, respectively, which are shown separately in Figure 6.1 as dashed lines. These are the expected error rates for an order-q representation of the current density and the same rates observed for several of the circular cylinder examples in Chapter 4. The error curves in Figure 6.1 should be compared with the corresponding error curves presented in Figure 4.6, obtained without the special singular treatment used in the corner cells.

Figure 6.2 shows a plot of the magnitude of the current density along half the perimeter of the triangle when the incident magnetic field has unity magnitude. Results from the LCN discretization of the TM MFIE are presented for a 78-cell model (26 cells/side) and $q = 3$, and compared to the reference data from [1] (markers on plot). The two results exhibit excellent agreement.

6.2   TM SCATTERING FROM A SQUARE CYLINDER

As a second example, consider a conducting cylinder with a square cross section. The interior angle at each corner is 90 degrees, for which the TM current density exponents from Equation (5.25) are

$$\left\{ -\frac{1}{3}, \ \frac{1}{3}, \ 1, \ \frac{5}{3}, \ \frac{7}{3}, \ 3, \ \frac{11}{3}, \ \dots \right\} . \qquad (6.3)$$

Table 5.1 from the preceding chapter summarized the desired representation as a function of order q when a 1:1 ratio of regular degrees of freedom (integer exponents) and singular degrees of

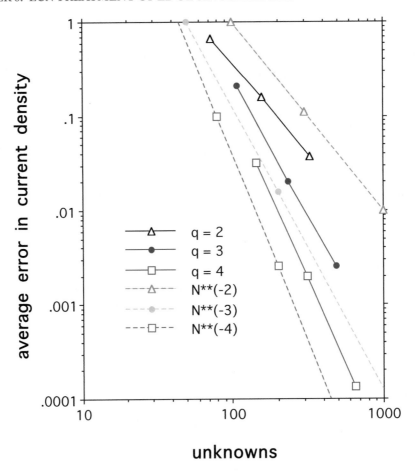

Figure 6.1: Error in the current density for a conducting cylinder whose cross section is an equilateral triangle. The cylinder has a perimeter dimension of 21 λ, and is illuminated by a TM plane wave incident symmetrically on one edge. Solid lines indicate the error in the TM MFIE solution averaged over four points located 1.5 λ, 2.0 λ, 2.5 λ, and 3.0 λ from the illuminated corner. Dashed lines show error rates of $O(h^2)$, $O(h^3)$, and $O(h^4)$ for comparison.

freedom (fractional exponents) are employed. Quadrature rules capable of exactly integrating these degrees of freedom are tabulated in Section B.7 of Appendix B, along with basis functions that realize the degrees of freedom from column 3 of Table 5.1.

The basic LCN discretization procedure follows in an identical fashion to the preceding example. The cylinder contour is divided into cells, with the corner cells twice as large as the others. The discretization procedure for non-corner cells is based on a Gauss-Legendre quadrature rule of

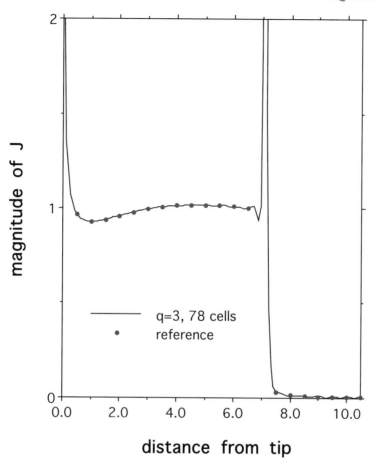

Figure 6.2: The TM current density induced on a conducting cylinder with an equilateral triangle cross section shape and a perimeter dimension of 21 λ. Results from the MFIE for $q = 3$ and a high order reference solution are shown for comparison.

order q, and the current density samples at the nodes of the rule are the unknowns (q per cell). In the cells adjacent to the corners of the square, we use the nodes of the $2q$-point generalized Gaussian quadrature rule from Section B.7 to define the current density samples ($2q$ unknowns per cell). For observer nodes closely spaced to source nodes, a new kernel L is synthesized either using Legendre polynomial basis functions (non-corner source cells) or the orthogonal singular basis functions from Section B.7 (corner cells). For observer nodes located a sufficient distance away from the source cell, a traditional Nyström sampling of the kernel K is used, with appropriate weights obtained from either the Gauss-Legendre rule (non-corner source cells) or the generalized rules (corner cells).

EFIE and MFIE implementations were developed and applied to a square cylinder of side dimension 5.25 λ illuminated by a uniform TM wave symmetrically incident upon one corner. The cylinder contour is represented by models consisting of 12, 19, 26, 33, or 40 cells per side of the cylinder, and an LCN discretization is carried out for q = 2 to 6. Figure 6.3 shows plots of the error in the MFIE current density obtained by averaging the error at 4 points on each face, with the points located 1.5 λ, 2.5 λ, 3.5 λ, and 4.5 λ from the leading corners. An error definition consistent with (3.22) was employed. Whenever those locations correspond to cell junctions, the interpolation procedure in (4.16)–(4.18) was used within both cells adjacent to each location to obtain the average current density at that location. A reference solution (Appendix C) obtained with a high order method-of-moments solution was used for comparison. The results in Figure 6.3 are converging faster than the expected $O(h^q)$ rates. While the reason for this superconvergence is not known, we note that a similar behavior was observed in connection with the TM MFIE in Chapter 4. Corresponding curves, not shown, were obtained for the TM EFIE, and those appear to converge at $O(h^q)$ rates for even values of q and $O(h^{q+1})$ rates for odd values of q. Results from both equations suggest that the singular representation is successful at producing high order behavior.

Figure 6.4 shows a plot of the magnitude of the current density along half the perimeter of the square target when the incident magnetic field has unity magnitude. Results from the LCN discretization of the TM EFIE are presented for q = 3 and a 160-cell model, and compared to the reference data from Appendix C (markers on plot). The two results exhibit excellent agreement.

6.3 TE SCATTERING FROM A SQUARE CYLINDER

As a third example, we reconsider the square conducting cylinder for the TE polarization. The interior angle at each corner is 90 degrees, for which the TE current density exponents from Equation (5.27) are

$$\left\{ 0, \ \frac{2}{3}, \ \frac{4}{3}, \ 2, \ \frac{8}{3}, \ \frac{10}{3}, \ 4, \ \dots \right\}. \tag{6.4}$$

As in the preceding sections, corner-cell representations will mix integer exponents with the non-integer exponents of (6.4), using twice as many degrees of freedom for a given order as the other cells. Section B.8 of Appendix B contains a tabulation of generalized Gaussian quadrature rules capable of integrating the appropriate degrees of freedom for the 90 degree TE wedge angle. Singular basis functions that implement these degrees of freedom and satisfy the orthogonality of (6.2) are also tabulated. Although the exponents in (6.4) are bounded, some of their derivatives are not, so we continue to refer to them as "singular" degrees of freedom.

The LCN discretization procedure follows in an identical fashion to the preceding examples, and we omit a further discussion of it. Here we consider the MFIE solution for a square cylinder of side dimension 5.25 λ illuminated by a uniform TE wave symmetrically incident upon one corner. The cylinder contour is represented by models consisting of 12, 19, 26, 33, or 40 cells per side of the cylinder, and an LCN discretization is carried out for q = 2 to 6. Figure 6.5 shows plots of the error in the MFIE current density obtained by averaging the error at 4 points on each face,

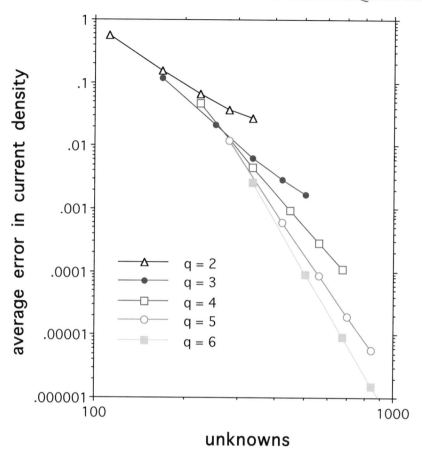

Figure 6.3: Error in the current density for a conducting cylinder having a square cross section shape. The cylinder has a perimeter dimension of 21 λ, and is illuminated by a TM plane wave incident symmetrically on one edge. Solid lines indicate the error in the TM MFIE solution averaged over 4 points on each face, with the points located 1.5 λ, 2.5 λ, 3.5 λ, and 4.5 λ from the leading corners.

with the points located 1.5 λ, 2.5 λ, 3.5 λ, and 4.5 λ from the leading corners. An error definition consistent with (3.22) was employed. A reference solution (Appendix C) obtained with a high order method-of-moments solution was used for comparison. These results closely follow the expected $O(h^q)$ rates.

Figure 6.6 shows a plot of the magnitude of the current density along half the perimeter of the square target, for $|H_z^{inc}| = 1$. Results from the LCN discretization of the TE MFIE are presented for $q = 4$ and a 104-cell model, and compared to the reference data from Appendix C (markers on plot). The two results exhibit excellent agreement.

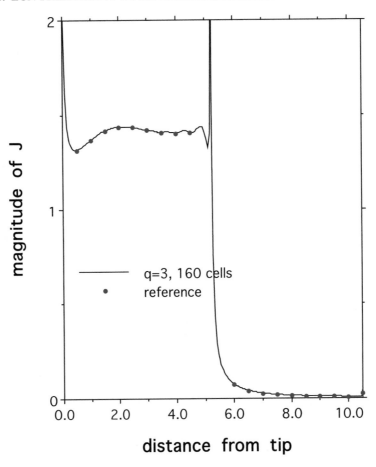

Figure 6.4: The TM current density induced along the surface of a conducting cylinder with a square cross section shape and a perimeter dimension of 21 λ (5.25 λ per side). Results from the EFIE for $q = 3$ and a high order reference solution (Appendix C) are shown for comparison.

Finally, Table 6.2 shows the two-dimensional scattering cross section, also known as the echo width, for the TM and TE case for the square cylinder of side dimension 5.25 λ. The scattering cross section is defined

$$\sigma_{TM} = \lim_{\rho \to \infty} 2\pi\rho \frac{\left|E_z^s(\phi)\right|^2}{\left|E_z^{inc}\right|_{target}^2} \tag{6.5}$$

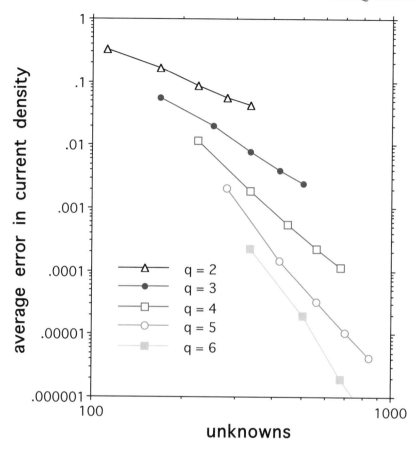

Figure 6.5: Error in the current density for a conducting cylinder having a square cross section shape. The cylinder has a perimeter dimension of 21 λ, and it is illuminated by a TE plane wave incident symmetrically on one edge. Solid lines indicate the error in the TE MFIE solution averaged over 4 points on each face, with the points located 1.5 λ, 2.5 λ, 3.5 λ, and 4.5 λ from the leading corners.

for the TM case and

$$\sigma_{TE} = \lim_{\rho \to \infty} 2\pi\rho \frac{\left| H_z^s(\phi) \right|^2}{\left| H_z^{\text{inc}} \right|^2_{\text{target}}} \tag{6.6}$$

for the TE case. The s superscript on the fields in (6.5) and (6.6) is used to denote the *scattered* fields, those produced by the current density on the target. Results in Table 6.2 have converged to the precision shown as the number of cells in the models and the representation order q are increased.

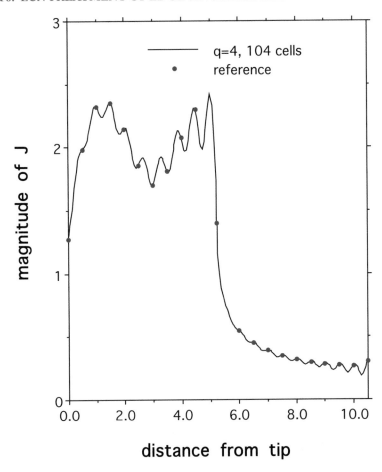

Figure 6.6: The TE current density induced along the surface of a conducting cylinder with a square cross section shape and a perimeter dimension of 21 λ (5.25 λ per side). Results from the MFIE for q = 4 and a high order reference solution (Appendix C) are shown for comparison.

6.4   INPUT IMPEDANCE OF A HOLLOW, LINEAR DIPOLE ANTENNA

There has been longstanding interest in the analysis of wire antennas. A special case of the general wire is the linear dipole. As our final example of an LCN implementation, we consider a hollow linear dipole of length $2h$ and radius a fed in the center by a magnetic frill source (in reality a model for a coaxial cable feeding a monopole of height h over a ground plane). Although this is physically a three-dimensional structure, due to the circular symmetry of the feed, the equation to be solved has a form very similar to that of the two-dimensional EFIE employed for TE scattering from cylinders

Table 6.2: Bistatic scattering cross section (echo width) of the square cylinder of side dimension 5.25 λ, in dB λ when illuminated by a plane wave incident symmetrically upon one edge. 180 degrees represents the backscattering direction.

angle (degrees)	σ_{TM}	σ_{TE}
0	25.398	25.572
15	0.580	1.595
30	2.508	2.731
45	1.520	−0.196
60	3.160	3.719
75	4.286	3.534
90	19.394	19.396
105	−0.089	−0.626
120	3.088	3.205
135	−0.847	−1.847
150	−5.717	−16.248
165	−2.971	−0.178
180	−2.446	1.056

in Section 4.5. The dipole EFIE can be written as

$$E_z^{\text{feed}}(z) = j\frac{\eta}{k}\left(\frac{d^2}{dz^2} + k^2\right)\int_{-h}^{h} I(z')G(z, z')\, dz' \tag{6.7}$$

where I is the total current, η is the intrinsic impedance, k is the wavenumber,

$$G(z, z') = \frac{1}{2\pi}\int_0^{2\pi} \frac{e^{-jkR}}{4\pi R}\, d\alpha \tag{6.8}$$

and

$$R = \sqrt{4a^2 \sin^2(\alpha/2) + (z - z')^2}\,. \tag{6.9}$$

We employ the exact dipole Green's function in (6.8), not the so-called reduced kernel. Equation (6.8) is evaluated by extracting the $1/R$ singular term and integrating it by means of a transformation to a complete elliptic integral. The remainder is evaluated by quadrature.

The excitation field from a magnetic frill source is given by

$$E_z^{\text{feed}}(\rho, z) = \frac{V_0}{\ln(b/a)}\int_{\phi'=0}^{2\pi}\left\{\frac{e^{-jkR_1}}{4\pi R_1} - \frac{e^{-jkR_2}}{4\pi R_2}\right\}\, d\phi' \tag{6.10}$$

where b is the outer radius of the coaxial feed, V_0 is the excitation voltage,

$$R_1 = \sqrt{\rho^2 + a^2 - 2\rho a \cos \phi' + z^2} \tag{6.11}$$

$$R_2 = \sqrt{\rho^2 + b^2 - 2\rho b \cos \phi' + z^2} . \tag{6.12}$$

At the ends of the dipole, the current has the same form as that occurring at a wedge angle of 0 degrees for the TE case. From Section B.2 of Appendix B, appropriate exponents for the current at the dipole ends are

$$\left\{ 0, \ \frac{1}{2}, \ 1, \ \frac{3}{2}, \ 2, \ \frac{5}{2}, \ 3, \ \frac{7}{2}, \ \dots \right\} . \tag{6.13}$$

Although the current is bounded at the ends, the surface charge density exhibits a singularity with a leading exponent of $-1/2$. Following the previous examples, we employ a representation where integer and fractional exponents are combined in a 1:1 ratio within the end cells. In this case, all the necessary terms are contained in the wedge solution and the basis functions (listed in Section B.2) follow the form of (6.13). As in the previous examples, the end cells are twice as large as the other nearby cells, with twice as many unknowns in those cells.

A fundamental difference between the dipole problem and the scattering problems considered previously is that the dipole feed field is highly localized. Thus, to improve the convergence of the input impedance, we consider the use of smaller cells in the immediate vicinity of the feed. Specifically, the closest cell to the feed was divided into 5 smaller cells, the next cell into 4, the next cell into 3, and so on. Beyond the fourth cell, they resume a uniform value up to the end cell, which is twice as large.

In the non-end cells, the LCN discretization proceeds using a Gauss-Legendre quadrature rule of order q, with the unknowns defined as the q current density samples at the nodes of the rule. For observer nodes located a sufficient distance away from the source cell, a traditional Nyström sampling of the integrand is carried out, with weights obtained from the q-point Gauss-Legendre rule. For observer nodes closely spaced (within 0.2λ or less in our implementation) to these cells, Legendre polynomial basis functions in the source cell are used to synthesize a new kernel L as previously described. An approach similar to that described in Equations (4.57)–(4.60) is used to evaluate the fields when the observer is located in the source cell. Equation (6.8) exhibits a logarithmic singularity as $z' \to z$, so lin-log rules are used within the local correction procedure.

For end cells, the alternate quadrature rules tabulated in Section B.2 of Appendix B are used to define the representation of $I(z)$, with the $2q$ samples at the nodes of the rule serving as the unknowns. For observer nodes located a sufficient distance away from the source cell, a traditional Nyström sampling of the integrand is used, with weights obtained from the alternate rules. For closely-spaced observer nodes, Equation (4.4) is solved to synthesize a new kernel using the set of basis functions given in Section B.2.

To illustrate the approach, consider a hollow dipole of length 0.5λ and radius 0.0625λ, driven by a frill of outer dimension $1.2a$. Table 6.3 presents results for input impedance for variable-cell models ranging from 34 to 82 cells, obtained by dividing V_0 in (6.10) by the current at the feed location, which is computed using the extrapolation process of Equations (4.16)–(4.18). These results appear to be converging to a value $Z_{in} \cong 84.81 - j\,29.68\,\Omega$. A high accuracy reference solution is not available for comparison, but we note that this value is within 4 Ω of that obtained by the authors in [5] for a solid dipole of these dimensions with flat end caps. Since the dipole is relatively thick, hollow and solid antennas should exhibit slightly different impedance values.

Table 6.3: Input impedance of a hollow linear dipole of length 0.50 λ and radius 0.0625 λ, in Ω. A frill feed is employed with $b/a = 1.2$.		
q, unknowns	Re$\{Z\}$	Im$\{Z\}$
3, 108	84.823	−29.676
3, 156	84.812	−29.682
3, 204	84.811	−29.682
5, 180	84.808	−29.680
5, 300	84.807	−29.680
5, 420	84.808	−29.681

As a second example, consider a dipole of length 0.47 λ and radius 0.005 λ, driven by a frill of outer dimension $2.23a$. Table 6.4 presents results for input impedance for variable-cell models, ranging from 50 to 114 cells. This is a thinner dipole than that considered above, and we note from [5] that more unknowns are typically required for comparable accuracy as the dipole radius is reduced. Results for $q=3$ and $q=5$ exhibit reasonable agreement with each other as the cell sizes are reduced, and to data presented in [6], which were obtained using an LCN approach employing a different implementation of the end condition. A high accuracy reference solution is not available for comparison. The results appear to be converging to a value $Z_{in} \cong 77.32 + j\,9.66\,\Omega$.

6.5 SUMMARY

In this chapter, several examples involving corner singularities in the current or charge density were illustrated. A representation that mixed integer exponents with fractional exponents arising from the appropriate wedge solution was employed in cells adjacent to an edge. The LCN implementation of this representation requires specialized Gaussian quadrature rules (Chapter 5 and Appendix B) for use in corner cells, as well as orthogonal singular basis functions for use within the local correction process. For the infinite triangular and square cylinder examples, for which reference data is available for comparison, error rates are observed to be the same or slightly better than those expected for smooth targets. This confirms that the singular representation used in corner cells is appropriate and that the LCN implementation produces high order behavior for targets with edges.

Table 6.4: Input impedance of a hollow linear dipole of length 0.47 λ and radius 0.005 λ, in Ω. A frill feed is employed with b/a = 2.23.

q, unknowns	Re$\{Z\}$	Im$\{Z\}$
3, 156	77.241	9.485
3, 204	77.293	9.591
3, 252	77.319	9.646
5, 420	77.321	9.655
5, 500	77.323	9.659
5, 580	77.323	9.659

REFERENCES

[1] M. M. Bibby, A. F. Peterson, and C. M. Coldwell, "High order representations for singular currents at corners," *IEEE Trans. Antennas Propagat.*, vol. 56, pp. 2277–2287, August 2008. DOI: 10.1109/TAP.2008.926771

[2] M. M. Bibby, A. F. Peterson, and C. M. Coldwell, "Optimum cell size for high order singular basis functions at geometric corners," *ACES Journal*, vol. 24, pp. 368–374, August 2009.

[3] S. Gedney, "Application of the high-order Nyström scheme to the integral equation solution of electromagnetic interaction problems, *Proceedings of the IEEE International Symposium on Electromagnetic Compatibility*, Washington, DC, pp. 289–294, August 2000. DOI: 10.1109/ISEMC.2000.875580

[4] M. S. Tong and W. C. Chew, "Nyström method with edge condition for electromagnetic scattering by 2D open structures," *Progress in Electromagnetic Research*, vol. 62, pp. 49–68, 2006. DOI: 10.2528/PIER06021901

[5] A. F. Peterson and M. M. Bibby, "High-order numerical solutions of the MFIE for the linear dipole," *IEEE Trans. Antennas Propagat.*, vol. 52, pp. 2684–2691, October 2004. DOI: 10.1109/TAP.2004.834407

[6] A. F. Peterson, "Application of the locally-corrected Nyström method to the EFIE for the linear dipole," *IEEE Trans. Antennas Propagat.*, vol. 52, pp. 603–605, February 2004. DOI: 10.1109/TAP.2004.823955

APPENDIX A

Parametric Description of Curved Cell Models

This Appendix describes the use of Bezier polynomials to represent curved cells in a parametric form. These allow curvature continuity at cell junctions and are easily incorporated into a Nyström or LCN code.

A.1 CUBIC BEZIER REPRESENTATION OF THE CYLINDER CONTOUR

As an alternative to the flat-cell approximation for the cylinder contour initially used in Chapter 3, consider a parameterization where the coordinates of points on each cell of the model are defined by a mapping of the form

$$x(u) = x_1 B_1(u) + x_2 B_2(u) + x_3 B_3(u) + x_4 B_4(u) \tag{A.1}$$
$$y(u) = y_1 B_1(u) + y_2 B_2(u) + y_3 B_3(u) + y_4 B_4(u) \tag{A.2}$$

where the basis functions are the Bernstein polynomials of cubic degree [1,2]

$$B_1(u) = (1 - u)^3 \tag{A.3}$$
$$B_2(u) = 3u(1 - u)^2 \tag{A.4}$$
$$B_3(u) = 3u^2(1 - u) \tag{A.5}$$
$$B_4(u) = u^3 \tag{A.6}$$

and where $0 < u < 1$. The four control points $\{x_i, y_i\}$ determine the position of the curve defined by (A.1)–(A.2). The first and last control point are the endpoints of the desired curve; the second and third control points control the shape of the curve but do not necessarily reside on the curve.

The mapping process will involve a transformation from the reference cell defined parametrically by $0 < u < 1$ to the curved domain in x-y space. Integrals over the curved cell can be evaluated in u-space and will involve the Jacobian of the transformation. The Jacobian is a function of the derivatives of the preceding polynomials, given by

$$B_1'(u) = -3(1 - u)^2 \tag{A.7}$$
$$B_2'(u) = 3(1 - 3u)(1 - u) \tag{A.8}$$
$$B_3'(u) = 3u(2 - 3u) \tag{A.9}$$
$$B_4'(u) = 3u^2 . \tag{A.10}$$

The derivatives of (A.1) and (A.2) can be obtained in terms of the control points as

$$\frac{\partial x}{\partial u} = 3(x_2 - x_1) + 6u(x_1 - 2x_2 + x_3) - 3u^2(x_1 - 3x_2 + 3x_3 - x_4) \tag{A.11}$$

$$\frac{\partial y}{\partial u} = 3(y_2 - y_1) + 6u(y_1 - 2y_2 + y_3) - 3u^2(y_1 - 3y_2 + 3y_3 - y_4). \tag{A.12}$$

The curved cell is defined by the position vector

$$\bar{r} = x(u)\,\hat{x} + y(u)\,\hat{y} \tag{A.13}$$

and a tangent vector

$$\bar{t} = \frac{\partial \bar{r}}{\partial u} = \frac{\partial x}{\partial u}\,\hat{x} + \frac{\partial y}{\partial u}\,\hat{y}. \tag{A.14}$$

The magnitude of the tangent vector is related to the Jacobian Q of the transformation (the arclength) by

$$Q(u) = |\bar{t}| = \sqrt{\left(\frac{\partial x}{\partial u}\right)^2 + \left(\frac{\partial y}{\partial u}\right)^2}. \tag{A.15}$$

A cubic mapping permits control over the tangent vector at the cell ends. At the ends of the interval,

$$\bar{t}\big|_{u=0} = 3(x_2 - x_1)\,\hat{x} + 3(y_2 - y_1)\,\hat{y} \tag{A.16}$$

$$\bar{t}\big|_{u=1} = 3(x_4 - x_3)\,\hat{x} + 3(y_4 - y_3)\,\hat{y}. \tag{A.17}$$

Thus, the two closest control points to one end of the cell determine the tangent vector at that end. An appropriate choice of these control points can, therefore, be used to maintain continuity of the tangent vector between cells.

The curvature of the curve at any location may be computed using [3, 4]

$$\kappa = \frac{\left|\bar{t} \times \frac{\partial \bar{t}}{\partial u}\right|}{|\bar{t}|^3} = \frac{\left|\left(\frac{\partial x}{\partial u}\right)\left(\frac{\partial^2 y}{\partial u^2}\right) - \left(\frac{\partial y}{\partial u}\right)\left(\frac{\partial^2 x}{\partial u^2}\right)\right|}{|\bar{t}|^3} \tag{A.18}$$

where, from above,

$$\frac{\partial^2 x}{\partial u^2} = 6(x_1 - 2x_2 + x_3) - 6u(x_1 - 3x_2 + 3x_3 - x_4) \tag{A.19}$$

$$\frac{\partial^2 y}{\partial u^2} = 6(y_1 - 2y_2 + y_3) - 6u(y_1 - 3y_2 + 3y_3 - y_4). \tag{A.20}$$

A.2   EXAMPLE: BEZIER MAPPING FOR CELLS ON A CIRCLE

As an example, suppose the preceding mapping functions are used to define the cells along a circular contour in such a way that the tangent vector is continuous from cell to cell. The task consists of defining the 4 control points (x_i, y_i) for a single cell. Consider a cell spanning an angle from $\phi = 0$ to $\phi = \phi_0$ on a circle of radius a. The two control points on the cell ends are defined by those endpoints, and therefore

$$x_1 = a, \quad y_1 = 0 \tag{A.21}$$
$$x_4 = a \cos \phi_0, \quad y_4 = a \sin \phi_0 . \tag{A.22}$$

The tangent vector along the cell at $\phi = 0$ is given by

$$\bar{t}\big|_{u=0} = 3(x_2 - a) \, \hat{x} + 3(y_2) \, \hat{y} . \tag{A.23}$$

For this to be tangential to the actual circle, it must have only a $\hat{y}$ component. Therefore,

$$x_2 = a . \tag{A.24}$$

A similar constraint may be applied to the tangent vector at $\phi = \phi_0$, which in that case leads to the constraint that (x_3, y_3) must lie on a line satisfying

$$x_3 \cos \phi_0 + y_3 \sin \phi_0 = a . \tag{A.25}$$

These constraints reduce the number of remaining parameters to two. By imposing symmetry on the control points, we eliminate one additional parameter and reduce their locations to the general form:

$$x_1 = a, \quad y_1 = 0 \tag{A.26}$$
$$x_2 = a, \quad y_2 = d \tag{A.27}$$
$$x_3 = a \cos \phi_0 + d \sin \phi_0, \quad y_3 = a \sin \phi_0 - d \cos \phi_0 \tag{A.28}$$
$$x_4 = a \cos \phi_0, \quad y_4 = a \sin \phi_0 \tag{A.29}$$

where d is the remaining parameter to be determined. We will select d in order to place the center of the parametric curve on the actual circle at radius a and angle $\phi_0/2$.

From the preceding equations, the center of the parametric curve segment is given by

$$x\left(\frac{1}{2}\right) = \frac{x_1 + 3x_2 + 3x_3 + x_4}{8} \tag{A.30}$$
$$y\left(\frac{1}{2}\right) = \frac{y_1 + 3y_2 + 3y_3 + y_4}{8} . \tag{A.31}$$

To place this on the circle, we must have

$$\left[x\left(\frac{1}{2}\right)\right]^2 + \left[y\left(\frac{1}{2}\right)\right]^2 = a^2 \tag{A.32}$$

which, after substitution, can be shown to be equivalent to

$$\frac{9}{32}d^2 \, (1 - \cos \phi_0) + \frac{9}{32}ad \sin \phi_0 - \frac{1}{2}a^2 \, (1 - \cos \phi_0) = 0 . \tag{A.33}$$

The relevant solution of this quadratic equation is

$$d = \frac{4}{3}a \left[\sqrt{1 + \left(\frac{\sin \phi_0}{1 - \cos \phi_0} \right)^2} - \left(\frac{\sin \phi_0}{1 - \cos \phi_0} \right) \right] . \qquad (A.34)$$

As an example, consider the case when $\phi_0 = 30°$. It follows from (A.34) that

$$d = 0.1755367a . \qquad (A.35)$$

Numerical values for the control points are

$$x_1 = a \qquad (A.36)$$
$$y_1 = 0 \qquad (A.37)$$
$$x_2 = a \qquad (A.38)$$
$$y_2 = d = 0.1755367a \qquad (A.39)$$
$$x_3 = a \cos \phi_0 + d \sin \phi_0 = 0.9537937a \qquad (A.40)$$
$$y_3 = a \sin \phi_0 - d \cos \phi_0 = 0.3479808a \qquad (A.41)$$
$$x_4 = \frac{\sqrt{3}}{2}a = 0.866025a \qquad (A.42)$$
$$y_4 = \frac{1}{2}a = 0.5a . \qquad (A.43)$$

With these values, we can determine the coordinates of the curved cell at any value of u using (A.1) and (A.2). The Jacobian in (A.15) is determined from (A.11) and (A.12), which specialize to

$$\frac{\partial x}{\partial u} = -0.277238au + 0.013933au^2 \qquad (A.44)$$

$$\frac{\partial y}{\partial u} = 0.526610a - 0.0185556au - 0.051997au^2 . \qquad (A.45)$$

For the $\phi_0 = 30°$ segment, the worst-case error in the radius is on the order of 0.00004%, and the maximum error in curvature is only about 0.01%.

Finally, note that the second and third control points are located at angles in ϕ given by:

$$\phi_2 = \tan^{-1} \left(\frac{0.1755367a}{a} \right) = 9.9561° \qquad (A.46)$$

$$\phi_3 = \tan^{-1} \left(\frac{0.3479808a}{0.9537937a} \right) = 20.0439° . \qquad (A.47)$$

Since the cell spans 30 degrees, the control points are very close to being uniformly spaced in angle. While the first and fourth control points are located on the desired circle, the others are some distance outside the circle.

A.3   JACOBIAN RELATIONSHIPS FOR THE INTEGRALS IN SECTION 4.5

To illustrate the use of parametric mappings, consider the integrals in Equation (4.56):

$$
\frac{4k}{\eta} I_{mj,nk} = k^2 \int_{t'=\alpha}^{\beta} \hat{t}(t_{mj}) \bullet \hat{t}(t') \, B_k(t') \, G(R_{mj}) dt'
$$
$$
+ \left\{ B_k(\alpha) \frac{\partial G}{\partial t} \Big|_{t=t_{mj}, t'=\alpha} - B_k(\beta) \frac{\partial G}{\partial t} \Big|_{t=t_{mj}, t'=\beta} \right\}
$$
$$
+ \frac{\partial B_k}{\partial t} \Big|_{t=t_{mj}} \left\{ G(t_{mj}, \alpha) - G(t_{mj}, \beta) \right\}
$$
$$
+ \int_{t'=\alpha}^{\beta} \left\{ \frac{\partial B_k}{\partial t'} \frac{\partial G}{\partial t} + \frac{\partial B_k}{\partial t} \frac{\partial G}{\partial t'} \right\} \Big|_{t=t_{mj}} dt' \tag{A.48}
$$

where $\{B_k\}$ denotes the basis function used for the local correction process and not the mapping function used above to define the curved cell shape. (In general, these are different, independent functions.)

Suppose that the cell shape is defined parametrically, as outlined above, by a mapping from a local cell $0 \le u \le 1$. Suppose also that the basis function and other quantities are directly defined in terms of the local variable, and the integral is to be performed over that domain. In other words, the quadrature rule used to evaluate (A.48) is now also defined over a domain $0 \le u \le 1$. The integrals can be evaluated using the Jacobian $Q(u)$ and the substitution

$$
dt' = Q(u')du' \tag{A.49}
$$

while the derivatives involve

$$
\frac{\partial}{\partial t'} = \frac{1}{Q(u')} \frac{\partial}{\partial u'} . \tag{A.50}
$$

The expression in (A.48) can, therefore, be recast as

$$
\frac{4k}{\eta} I_{mj,nk} = k^2 \int_{u'=0}^{1} \hat{t}(u_{mj}) \bullet \hat{t}(u') \, B_k(u') \, G(R_{mj}) Q(u') du'
$$
$$
+ \left\{ B_k(0) \frac{1}{Q(u_{mj})} \frac{\partial G}{\partial u} \Big|_{u=u_{mj}, u'=0} - B_k(1) \frac{1}{Q(u_{mj})} \frac{\partial G}{\partial u} \Big|_{u=u_{mj}, u'=1} \right\}
$$
$$
+ \frac{1}{Q(u_{mj})} \frac{\partial B_k}{\partial u} \Big|_{u=u_{mj}} \left\{ G(u_{mj}, 0) - G(u_{mj}, 1) \right\}
$$
$$
+ \int_{u'=0}^{1} \frac{1}{Q(u_{mj})} \left\{ \frac{\partial B_k}{\partial u'} \frac{\partial G}{\partial u} + \frac{\partial B_k}{\partial u} \frac{\partial G}{\partial u'} \right\} \Big|_{u=u_{mj}} du' \tag{A.51}
$$

where the variables are now evaluated at the locations specified by u and u'. Furthermore, the other Nyström matrix entries in (4.44) are obtained as

$$
Z_{mj,ni} = w_i \, Q(u_{ni}) D(u_{mj}, u_{ni}) \tag{A.52}
$$

where it is assumed that the Nyström quadrature rule is also defined over $0 \leq u \leq 1$.

REFERENCES

[1] D. F. Rogers, *An Introduction to NURBS with Historical Perspective*. San Francisco: Morgan Kaufmann, 2001, pp. 17-35.

[2] L. Piegl and W. Tiller, *The NURBS Book*. Berlin: Springer-Verlag, 1997, pp. 9-24.

[3] A. E. Taylor, *Advanced Calculus*. Boston: Ginn and Company, 1955, pp. 366-369.

[4] B. O'Neill, *Elementary Differential Geometry*. New York: Academic Press, 1966, pp. 56-69.

APPENDIX B

Basis Functions and Quadrature Rules for Edge Cells

The following tables present basis functions for use in cells adjacent to an edge and quadrature rules that can exactly integrate those degrees of freedom. The basis functions are defined on the domain $0 \le u \le 1$ and contain terms having a singularity at $u = 0$. These satisfy the orthogonality condition

$$\int_0^1 B_i(u)B_j(u)du = 0, \quad i \ne j .$$

B.1 TM CASE, WEDGE ANGLE $= 0$ DEGREES

The exponents associated with a wedge angle of 0 degrees, for the TM polarization, are

$$\left\{ -\frac{1}{2}, \ 0, \ \frac{1}{2}, \ 1, \ \frac{3}{2}, \ 2, \ \frac{5}{2}, \ 3, \ \frac{7}{2}, \ \dots \right\} .$$

Due to the need to apply the Gram-Schmidt orthogonalization process to functions containing the $u^{-1/2}$ singularity, which is not integrable if multiplied with itself, that term is placed at the end of the list. Thus, for this case only, we group the basis functions independently for each order q:

For $q=1$, wedge angle = 0, TM:

$$B_1 = 1$$
$$B_2 = 1 - \frac{1}{2}u^{-1/2}$$

For $q=2$, wedge angle = 0, TM:

$$B_1 = u$$
$$B_2 = u - \frac{5}{6}u^{1/2}$$
$$B_3 = u - \frac{4}{3}u^{1/2} + \frac{2}{5}$$
$$B_4 = u - \frac{3}{2}u^{1/2} + \frac{3}{5} - \frac{1}{20}u^{-1/2}$$

For $q=3$, wedge angle $= 0$, TM:

$$B_1 = u^2$$

$$B_2 = u^2 - \frac{9}{10}u^{3/2}$$

$$B_3 = u^2 - \frac{8}{5}u^{3/2} + \frac{28}{45}u$$

$$B_4 = u^2 - \frac{21}{10}u^{3/2} + \frac{7}{5}u - \frac{7}{24}u^{1/2}$$

$$B_5 = u^2 - \frac{12}{5}u^{3/2} + 2u - \frac{2}{3}u^{1/2} + \frac{1}{14}$$

$$B_6 = u^2 - \frac{5}{2}u^{3/2} + \frac{20}{9}u - \frac{5}{6}u^{1/2} + \frac{25}{210} - \frac{1}{252}u^{-1/2}$$

Weights and nodes associated with the generalized quadrature rule for the TM case, wedge angle $= 0$.

nodes	u_i	w_i
2	0.04465819873852045108	0.21132486540518711775
(q=1)	0.62200846792814621559	0.78867513459481288225
4	0.00482078098942601431	0.02415220341283324407
(q=2)	0.10890625570683385139	0.21521408227178500202
	0.44888729929169011619	0.43693107259076114061
	0.86595709258347858953	0.32370264172462061330
6	0.00114009162798957143	0.00578481309963187582
(q=3)	0.02869476995464122862	0.06111131733618720551
	0.14492518595015338399	0.17813034617398462949
	0.38354437203335029262	0.28978358839870641790
	0.68990415642090574228	0.29965025571195140206
	0.93360960583114159924	0.16553967927953846922

B.2   TE CASE, WEDGE ANGLE = 0 DEGREES

Wedge angle exponents: $\left\{0, \; \frac{1}{2}, \; 1, \; \frac{3}{2}, \; 2, \; \frac{5}{2}, \; 3, \; \frac{7}{2}, \; \ldots\right\}$

For $q=1$, 2, or 3, wedge angle = 0, TE:

$$B_1 = 1$$

$$B_2 = 1 - \frac{3}{2}u^{1/2}$$

$$B_3 = 1 - 4u^{1/2} + \frac{10}{3}u$$

$$B_4 = 1 - \frac{15}{2}u^{1/2} + 15u - \frac{35}{4}u^{3/2}$$

$$B_5 = 1 - 12u^{1/2} + 42u - 56u^{3/2} + \frac{126}{5}u^2$$

$$B_6 = 1 - \frac{35}{2}u^{1/2} + \frac{280}{3}u - 210u^{3/2} + 210u^2 - 77u^{5/2}$$

Weights and nodes associated with the generalized quadrature rule for the TE case, wedge angle = 0.

nodes	u_i	w_i
2	0.12245873770769437179	0.35425685251863018197
(q=1)	0.70712040042119297903	0.64574314748136981803
4	0.01524856844722977212	0.04994558737747997161
(q=2)	0.14941996653468718390	0.23888804335613144116
	0.49161039093402031768	0.41683394529516138341
	0.87867947254287240495	0.29433242397122720383
6	0.00306214692356179302	0.01045180879247385085
(q=3)	0.03585950335748828446	0.06512282379833040652
	0.15268928349594617396	0.17533030465608331172
	0.38686165306057465181	0.28494377475304237368
	0.68984841648072294991	0.29808548879196502557
	0.93332909547871415974	0.16606579920810503166

B.3   TM CASE, WEDGE ANGLE $= 30$ DEGREES

Wedge angle exponents: $\left\{-\frac{5}{11}, \frac{1}{11}, \frac{7}{11}, \frac{13}{11}, \frac{17}{11}, \frac{19}{11}, \frac{23}{11}, \cdots\right\}$

For q=1, 2, or 3, wedge angle = 30, TM:

$B_1 = u^{-5/11}$

$B_2 = u^{-5/11} - 6$

$B_3 = u^{-5/11} - 66 + 70u^{1/11}$

$B_4 = u^{-5/11} - \frac{1152}{7} + 208u^{1/11} - \frac{391}{7}u$

$B_5 = u^{-5/11} - \frac{304128}{833} + \frac{9568}{17}u^{1/11} + \frac{2277}{7}u - \frac{429780}{833}u^{7/11}$

$B_6 = u^{-5/11} - \frac{6718464}{10829} + \frac{125856}{119}u^{1/11} + \frac{162081}{91}u - \frac{1487700}{833}u^{7/11} - \frac{40960}{91}u^2$

Weights and nodes associated with the generalized quadrature rule for the TM case, wedge angle $= 30$.

nodes	u_i	w_i
2	0.03679086462548340237	0.17728648468311316408
(q=1)	0.59981690801812300936	0.82271351531688683592
4	0.00220779650623542903	0.01170288339174060310
(q=2)	0.07348474047135435107	0.17061367844990569407
	0.39136878362869437671	0.44945791344532918601
	0.84603820959609780583	0.36822552471302451682
6	0.00028098822081717140	0.00153453277237272824
(q=3)	0.01175050400070804195	0.03071495346689777286
	0.08875104579225131045	0.13777595209688510595
	0.30343569535918258754	0.28862219218388499196
	0.63263448173425167638	0.34122501294754631248
	0.91922419394528172875	0.20012735653241308852

B.4   TE CASE, WEDGE ANGLE = 30 DEGREES

Wedge angle exponents: $\left\{ \frac{6}{11}, \frac{12}{11}, \frac{18}{11}, \frac{24}{11}, \frac{28}{11}, \frac{30}{11}, \cdots \right\}$

For $q=1, 2,$ or 3, wedge angle = 30, TE:

$$B_1 = u^{6/11}$$

$$B_2 = u^{6/11} - \frac{17}{23}$$

$$B_3 = u^{6/11} - \frac{11}{46} + \frac{29}{34} u^{12/11}$$

$$B_4 = u^{6/11} - \frac{5}{58} + \frac{175}{34} u^{12/11} - \frac{4032}{667} u$$

$$B_5 = u^{6/11} - \frac{55}{1218} + \frac{25}{2} u^{12/11} + \frac{57024}{4669} u - \frac{10250}{8211} u^{18/11}$$

$$B_6 = u^{6/11} - \frac{4}{147} + \frac{164}{7} u^{12/11} - \frac{663552}{32683} u - \frac{385400}{57477} u^{18/11} + \frac{85293}{32683} u^2$$

Weights and nodes associated with the generalized quadrature rule for the TE case, wedge angle = 30.

nodes	u_i	w_i
2	0.11703992040309433787	0.34443905947501908963
(q=1)	0.70121151441268668316	0.65556094052498091037
4	0.01364516033212367046	0.04563410475615564479
(q=2)	0.14159535956755051933	0.23267708034971305858
	0.48186337690431127366	0.42022161496575359213
	0.87552448191998878826	0.30146719992837770450
6	0.00258276409397914729	0.00901782889524804101
(q=3)	0.03235585618123590874	0.06055311226699369263
	0.14416764456625731955	0.17079705824621701139
	0.37610910203365405481	0.28558874743916820920
	0.68251797883693458860	0.30360715343948762902
	0.93151734807661131734	0.17043609971288541675

B.5   TM CASE, WEDGE ANGLE $= 60$ DEGREES

Wedge angle exponents: $\left\{ \frac{-2}{5}, \ \frac{1}{5}, \ \frac{4}{5}, \ \frac{7}{5}, \ \frac{8}{5}, \ 2, \ \frac{11}{5}, \ \frac{13}{5}, \ \ldots \right\}$

For q=1, 2, or 3, wedge angle = 60, TM:

$$B_1 = u^{-2/5}$$
$$B_2 = u^{-2/5} - 3$$
$$B_3 = u^{-2/5} - 15 + 16u^{1/5}$$
$$B_4 = u^{-2/5} - \frac{63}{2} + 49u^{1/5} - 22u$$
$$B_5 = u^{-2/5} - \frac{945}{16} + \frac{539}{4}u^{1/5} + \frac{495}{2}u - \frac{5145}{16}u^{4/5}$$
$$B_6 = u^{-2/5} - \frac{729}{8} + \frac{770}{3}u^{1/5} + 1188u - \frac{9555}{8}u^{4/5} - \frac{494}{3}u^2$$

Weights and nodes associated with the generalized quadrature rule for the TM case, wedge angle = 60.

nodes	u_i	w_i
2	0.04311039568060804414	0.19516532986695269323
(q=1)	0.61079171120328984626	0.80483467013304730677
4	0.00302298209383438970	0.01500619801024183126
(q=2)	0.82425878234313919033	0.18150875423224871565
	0.40549844784519385506	0.44603627261337513561
	0.85082569301656332611	0.35744877514413431748
6	0.00430354935356894482	0.00219985637203055060
(q=3)	0.01446824047058494474	0.03574488250064748638
	0.09829080395178143985	0.14505684125410774146
	0.31741178614734506409	0.28897626565299042962
	0.64263581459190139635	0.33391027367555150214
	0.92172081543146628585	0.19411188054467228980

B.6 TE CASE, WEDGE ANGLE = 60 DEGREES

Wedge angle exponents: $\left\{0, \ \frac{3}{5}, \ \frac{6}{5}, \ \frac{9}{5}, \ \frac{12}{5}, \ \frac{13}{5}, \ 3, \ \ldots\right\}$

For $q=1, 2$, or 3, wedge angle = 60, TE:

$$B_1 = u^{3/5}$$

$$B_2 = u^{3/5} - \frac{8}{11}$$

$$B_3 = u^{3/5} - \frac{5}{22} - \frac{7}{8}u^{6/5}$$

$$B_4 = u^{3/5} - \frac{1}{14} + \frac{17}{8}u^{6/5} - \frac{234}{77}u$$

$$B_5 = u^{3/5} - \frac{10}{273} + \frac{68}{13}u^{6/5} - \frac{405}{77}u - \frac{1615}{1716}u^{9/5}$$

$$B_6 = u^{3/5} - \frac{14}{663} + \frac{140}{13}u^{6/5} - \frac{1539}{187}u - \frac{15295}{1716}u^{9/5} + \frac{13122}{2431}u^2$$

Weights and nodes associated with the generalized quadrature rule for the TE case, wedge angle = 60.

nodes	u_i	w_i
2	0.12245873770769437179	0.35425685251863018197
(q=1)	0.70712040042119297903	0.64574314748136981802
4	0.01524856844722977212	0.04994558737747997161
(q=2)	0.14941996653468718390	0.23888804335613144116
	0.49161039093402031768	0.41683394529516138341
	0.87867947254287240495	0.29433242397122720383
6	0.00306214692356179302	0.01045180879247385085
(q=3)	0.03585950335748828446	0.06512282379833040652
	0.15268928349594617396	0.17533030465608331172
	0.38686165306057465181	0.28494377475304237368
	0.68984841648072294991	0.29808548879196502557
	0.93332909547871415974	0.16606579920810503166

B.7 TM CASE, WEDGE ANGLE $= 90$ DEGREES

Wedge angle exponents: $\left\{ -\frac{1}{3},\ \frac{1}{3},\ 1,\ \frac{5}{3},\ \frac{7}{3},\ 3,\ \frac{11}{3},\ \dots \right\}$

For q=1, 2, or 3, wedge angle = 90, TM:

$$B_1 = u^{-1/3}$$

$$B_2 = u^{-1/3} - 2$$

$$B_3 = u^{-1/3} - 6 + 6u^{1/3}$$

$$B_4 = u^{-1/3} - \frac{32}{3} + 20u^{1/3} - \frac{35}{3}u$$

$$B_5 = u^{-1/3} - \frac{384}{25} + 42u^{1/3} - 63u + \frac{924}{25}u^{5/3}$$

$$B_6 = u^{-1/3} - \frac{512}{25} + \frac{378}{5}u^{1/3} - 231u + \frac{12012}{25}u^{5/3} - \frac{1536}{5}u^2$$

Weights and nodes associated with the generalized quadrature rule for the TM case, wedge angle = 90.

nodes	u_i	w_i
2	0.50857653817424135321	0.21549834031446860964
(q=1)	0.62337696035734301197	0.78450165968553139036
4	0.00565599779924914190	0.02553902985876776627
(q=2)	0.11059220815356055150	0.21547389046137733420
	0.44996320074510315157	0.43577791410681540233
	0.86613197498872463885	0.32320916557303949720
6	0.00134764232860906757	0.00617377107250384220
(q=3)	0.02930714197595218078	0.06152435565779431847
	0.14576352329062226218	0.17811939885163736022
	0.38416983624086738666	0.28942788917886168743
	0.69016603351648062557	0.29933378271366836635
	0.93365398508033003356	0.16542080252553442533

B.8   TE CASE, WEDGE ANGLE = 90 DEGREES

Wedge angle exponents: $\left\{0, \frac{2}{3}, \frac{4}{3}, 2, \frac{8}{3}, \frac{10}{3}, 4, \ldots\right\}$

For q=1, 2, or 3, wedge angle = 90, TE:

$$B_1 = u^{2/3}$$

$$B_2 = u^{2/3} - \frac{5}{7}$$

$$B_3 = u^{2/3} - \frac{3}{14} - \frac{9}{10}u^{4/3}$$

$$B_4 = u^{2/3} - \frac{1}{18} + \frac{11}{10}u^{4/3} - \frac{128}{63}u$$

$$B_5 = u^{2/3} - \frac{1}{32} + \frac{33}{16}u^{4/3} - \frac{96}{35}u - \frac{143}{480}u^{8/3}$$

$$B_6 = u^{2/3} - \frac{11}{624} + \frac{495}{104}u^{4/3} - \frac{256}{65}u + \frac{209}{240}u^{8/3} - \frac{187}{70}u^2$$

Weights and nodes associated with the generalized quadrature rule for the TE case, wedge angle = 90.

nodes	u_i	w_i
2	0.12891217317815423349	0.36578103754701189353
(q=1)	0.71402212540439801012	0.63421896245298810647
4	0.02015969421104826999	0.06337480404408955196
(q=2)	0.17433499913864280578	0.25938254606961172736
	0.52330277009994161425	0.40629112545108640538
	0.88905074983693049596	0.27095152443521231529
6	0.00552522483166338268	0.01782614051965890468
(q=3)	0.05369623484637024041	0.08805674117653229339
	0.19522466490311081812	0.19719739650042051254
	0.43983909710906238391	0.28134707231302751648
	0.72582756994400026920	0.27096106060294532888
	0.94222701016667326229	0.14461158888741544402

APPENDIX C

Reference Data for Square Cylinder

The following tables present reference data for the current density induced on a perfectly conducting square cylinder of side dimension 5.25 wavelengths. The results were obtained from high order method of moments (MoM) discretizations of the EFIE and MFIE, based on the procedure in [1] for treating the current representation near corners, and is tabulated here for comparison with the LCN procedure of Chapter 6.

REFERENCES

[1] M.M. Bibby, A.F. Peterson, and C.M. Coldwell, "Optimum cell size for high order singular basis functions at geometric corners," *ACES Journal*, vol. 24, pp. 368-374, August 2009.

Table C.1: Currents induced on a cylinder whose cross section is a square with side dimension 5.25 λ. A TM-to-z uniform plane wave is incident symmetrically upon one corner, with the incident magnetic field normalized to unity, and the phase reference at the center of the square. Distance is measured from the illuminated corner along one side and across the back side of the cylinder to the opposite corner. To the precision shown, the MoM results did not change for q = 6 to q = 8. The model used cells of 0.25 λ except at the corners, where they were 0.5 λ (19 cells per side) in accordance with the approach of [1]. Phase is in degrees.

TM EFIE or TM MFIE MoM Results		
Distance	Amplitude	Phase
0.001	8.47034	46.405
0.500	1.31129	-53.965
1.000	1.36349	-176.863
1.500	1.41456	56.066
2.000	1.43549	-72.130
2.500	1.43204	159.743
3.000	1.41782	32.142
3.500	1.40541	-94.942
4.000	1.40121	138.227
4.500	1.40479	11.422
5.249	4.32518	151.912
5.251	4.17439	148.387
6.000	0.07494	172.560
6.500	0.04141	-16.799
7.000	0.02708	157.568
7.500	0.01942	-26.246
8.000	0.01476	150.975
8.500	0.01169	-31.182
9.000	0.00953	147.003
9.500	0.00795	-34.818
10.000	0.00675	142.118
10.499	0.03329	-65.252

Table C.2: Currents induced on a cylinder whose cross section is a square with side dimension 5.25 λ. A TE-to-z uniform plane wave is incident symmetrically upon one corner, with the incident magnetic field normalized to unity, and the phase reference at the center of the square. Distance is measured from the illuminated corner along first the lit side and across the shadow side of the cylinder to the opposite corner. The MFIE results for $q = 6$ to $q = 8$, and the EFIE results for $q = 7$ to $q = 8$, only differed in the last digit shown below. The model used cells of 0.25 λ except at the corners, where they were 0.5 λ (19 cells per side) in accordance with the approach of [1]. Phase is in degrees.

	TE EFIE or TE MFIE MoM Results	
Distance	Amplitude	Phase
0.000	1.26729	78.559
0.500	1.98288	-36.612
1.000	2.32295	-171.508
1.500	2.35182	52.440
2.000	2.14561	-80.965
2.500	1.85380	151.729
3.000	1.69646	31.829
3.500	1.80993	-87.433
4.000	2.07618	146.633
4.500	2.29939	14.877
5.250	1.39578	176.650
5.250	1.39578	176.650
6.000	0.54759	-127.491
6.500	0.44799	49.077
7.000	0.38906	-132.973
7.500	0.34925	45.595
8.000	0.32042	-135.521
8.500	0.29878	43.521
9.000	0.28253	-137.388
9.500	0.27130	41.650
10.000	0.26765	-139.430
10.500	0.30045	44.660

Authors' Biographies

ANDREW F. PETERSON

Andrew F. Peterson received the B.S., M.S., and Ph.D. degrees in Electrical Engineering from the University of Illinois, Urbana-Champaign in 1982, 1983, and 1986 respectively. Since 1989, he has been a member of the faculty of the School of Electrical and Computer Engineering at the Georgia Institute of Technology, where he is now Professor and Associate Chair for Faculty Development. He teaches electromagnetic field theory and computational electromagnetics, and conducts research in the development of computational techniques for electromagnetic scattering, microwave devices, and electronic packaging applications.

MALCOLM M. BIBBY

Malcolm M. Bibby received the B.Eng. and Ph.D. degrees in Electrical Engineering from the University of Liverpool, England in 1962 and 1965, respectively. He also holds an MBA from the University of Chicago, U.S.A. His career includes both engineering and management. He was president of LXE Inc., a manufacturer of wireless data communications products from 1983 to 1994. Thereafter he was president of NDI, a manufacturer of hardened hand-held computers, for five years. He is currently an Adjunct Professor in ECE at Georgia Tech. He has been interested in the numerical aspects of antenna design for the last twenty-five years.

Printed in the United States
by Baker & Taylor Publisher Services